AGRICULTURAL DRAINAGE ENGINEERING

AGRICULTURAL DRAINAGE ENGINEERING

by
Aishwarya Malik | Tarun Chaudhary | Nikhil Bhagat

Wordpen Academics
Publishing Science That Matter

ISBN: 978-93-6084-058-7 (Hardback)

Published by : **WORDPEN ACADEMICS**
4736/23, Ansari Road, Darya Ganj,
New Delhi – 110 002
Phone: +91-011-35004336
E-mail: info@wordpenacademics.com

Printed at **:** **Replika Press Pvt. Ltd.**

Exclusive Distributor: ***Bio-Green Books, New Delhi***

Preface

"Agricultural Drainage Engineering" serves as a definitive guide in understanding and mastering the intricacies of managing water in agricultural landscapes. This book is a testament to the critical role that drainage systems play in enhancing agricultural productivity, sustainability, and resilience in diverse agro-ecosystems worldwide.

Water, often considered the lifeblood of agriculture, can be both a precious resource and a formidable challenge for farmers and land managers. Effective drainage engineering is essential for mitigating water-related risks such as waterlogging, salinity, erosion, and nutrient loss, while simultaneously optimizing soil moisture conditions for crop growth and yield.

As authors of this volume, we have endeavored to compile a comprehensive resource that integrates theoretical foundations with practical applications, drawing upon the collective expertise of leading scholars, researchers, and practitioners in the field of agricultural drainage engineering.

This book traverses a wide spectrum of topics, ranging from the principles of hydrology and soil-water-plant interactions to the design, implementation, and management of drainage systems tailored to diverse agricultural contexts. Through a multidisciplinary approach, readers will gain insights into the latest advancements in drainage technologies, modeling techniques, decision support systems, and sustainable drainage practices.

Our overarching goal in presenting this book is to bridge the gap between theory and practice, providing both students and professionals with the requisite knowledge and tools to address the complex water management challenges confronting modern agriculture.

We extend our sincere gratitude to all the contributors who have generously shared their expertise, experiences, and innovations. It is our hope that this volume will serve as a valuable resource for researchers, educators, policymakers, and practitioners alike, fostering a deeper understanding and appreciation for the pivotal role of agricultural drainage engineering in shaping the future of sustainable agriculture.

Aishwarya Malik | Tarun Chaudhary | Nikhil Bhagat

Contents

1

Agricultural Drainage System

An agricultural drainage system is a system by which the water level on or in the soil is controlled to enhance agricultural crop production

Agricultural drainage is the removal of excess water from the soil surface and/ or soil profile of cropland, by either gravity or artificial means. The two main reasons for improving the drainage on agricultural land are for soil conservation and enhancing crop production. Research conducted in Ohio and throughout the Midwest has documented many benefits of agricultural drainage improvement.

Drainage is of two forms :

1. Surface drainage; and
2. Sub surface drainage or underground drainage.

Surface drainage (Natural system of drainage) : It may consist of open ditches that are laid out by eye judgment, leading from one wet spot to another and finally into a nala or river. This is often called natural system.

Advantages of Drainage Systems

(i) Removes toxic material and disease organisms. Also, reduces soil erosion by creating better root systems.
(ii) Checks planting delays resulting in a longer growing season.
(iii) Reduces the effect of droughts and excessive water.
(iv) Improves aeration and the nitrogen content.

Open ditch drains : The pattern of ditches is regular. The method is adopted to land that has uniform slope.

Field ditches : Field ditches for surface drains may be either narrow with nearly vertical sides or Vshaped with flat side slopes. V-shaped ditches have the advantages of being easier to cross with large machinery.

Narrow ditches : These are most common where large farm machinery is not used.

In level areas, a collecting ditch may need to be installed at one side of the field and shallow shaped ditches are constructed to discharge into the collecting ditch. The field ditches should be laid out parallel and spaced 15 to 45 meters or more apart as required by the soil surface conditions and crop to be grown. They should be 30 to 60 cm deep depending upon the depth of the collecting ditch.

Farming operations should be parallel to the field ditches. The care that a ditch will drain satisfactorily depends up on how quickly water runs into the ditch how much rain falls on the land, slope, and the condition of the soil and plant cover.

Sub surface or under ground drainage : A sub surface or underground drainage will remove excess soil water. It percolates in to themselves, just like open drains. These underground drains afford the utility that the surface of the field is not cut off, no wastage of lad and do not interfere with farm operations. On the other hand, they are costly to lie and are useless in slowly permeable clay soils.

Underground drains may be classified as:

1. *Tile drain*: It consists of digging a narrow trench, placing short section of tiles at the bottom and covering the tiles with earth. The loose joints between two section of the tiles serve as a opening where drainage water may enter into the drainage system. Water moves by gravity into the joins between tiles and through tile walls.

   Porous tile gives no better drainage than tiles that water does not percolate and porous tile can easily broken or crushed. the drains are two types of tiles in use. Tile should be always placed at least 75 cm deep to prevent breakage by heavy machinery.
2. *Box drains*: Instead of pipes, underground drains may be made in V-shaped cut or trench, sides of which are reverted with soil, restoring the surface of the field. Depth may be 90 cm below ground.
3. *Rubble drains*: A somewhat equally substitute for tile drains is made by cutting narrow V-shaped drains or rectangular in section, as for box drains, filling them up with rough stones large and small and then covering the whole up with soil level with surface field soil. Depth may be 90 cm.
4. *Mole drains*: They are often used in clay, clay loam soils. A moling machine is one that draws a bullet nosed cylinder; usually 10-15 cm in diameter is therefore formed. A mole drain should be at least 75 cm below the surface to prevent closing of the holes by compaction from farming operations. Mole drains are extremely used in Europe.
5. *Drainage pipe applications* : The pumps are used in U.S.A. and many other countries for drainage. River bottoms, lakes and costal plains, peat lands and irrigated lands are the main types of lands reclaimed by pump drainage. The subsequent must be sufficiently permeable for the ground water to move to the pipes enough for effective pumping.

Function

The function of the field drainage system is to control the water table, whereas the function of the main drainage system is to collect, transport, and dispose of the water through an outfall or outlet. In some instances one makes an additional distinction between collector and main drainage systems.

Field drainage systems are differentiated in surface and subsurface field drainage systems.

Sometimes like in irrigated, submerged rice fields, a form of temporary drainage is required whereby the drainage system is allowed to function on certain occasions only (e.g. during the harvest period). If allowed to function continuously, excessive quantities of water would be lost. Such a system is therefore called a checked, or controlled, drainage system.

However, the drainage system is meant to function as regularly as possible to prevent undue water logging at any time and one employs a regular drainage system. In literature, this is sometimes also called a "relief drainage system".

The use of surface and subsurface drainage improvements is not limited to agricultural lands. Many residential homes use subsurface drainage systems, similar to those used in agriculture, to prevent water damage to foundations and basements. Golf courses make extensive use of both surface and subsurface drains. Houses, streets and buildings in urban areas depend heavily on surface and subsurface drainage systems for protection. These generally are a combination of plastic or metal gutters, and concrete pipes or channels.

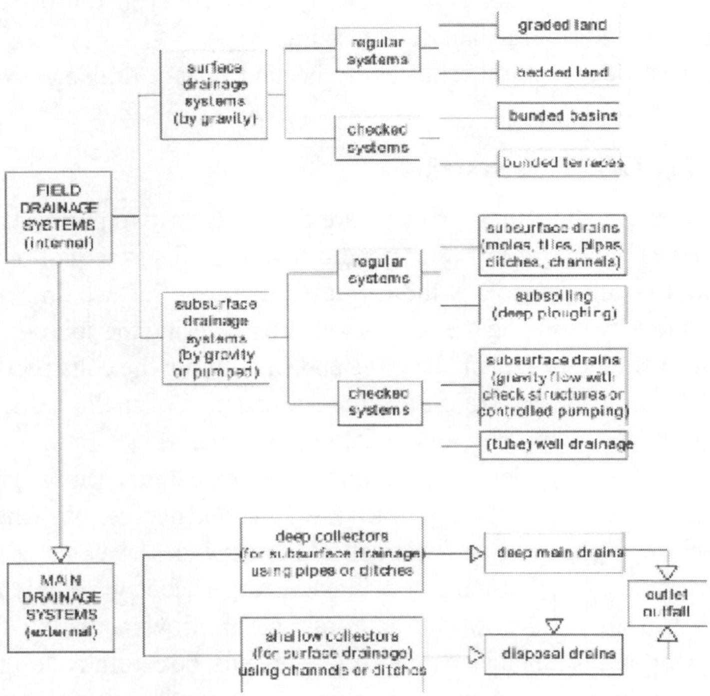

Classification of agricultural drainage systems

The above figure classifies the various types of drainage systems. It shows the field (or internal) and the main (or external) systems.

SURFACE DRAINAGE SYSTEMS

The regular surface drainage systems, which start functioning as soon as there is an excess of rainfall or irrigation, operate entirely by gravity. They consist of reshaped or reformed land surfaces and can be divided into:

- Bedded systems : It is used in flat lands for crops other than rice; and
- Graded systems : It is used in sloping land for crops other than rice.

The bedded and graded systems may have ridges and furrows. The checked surface drainage systems consist of check gates placed in the embankments surrounding flat basins, such as those used for rice fields in flat lands. These fields are usually submerged and only need to be drained on certain occasions like at harvest time. Checked surface drainage systems are also found in terraced lands used for rice.

In literature, not much information can be found on the relations between the various regular surface field drainage systems, the reduction in the degree of waterlogging, and the agricultural or environmental effects. It is therefore difficult to develop sound agricultural criteria for the regular surface field drainage systems. Most of the known criteria for these systems concern the efficiency of the techniques of land leveling and earthmoving.

Similarly, agricultural criteria for checked surface drainage systems are not very well known.

Subsurface Drainage Systems

Surface drainage improvements are designed for two purposes: to minimize crop damage resulting from water ponding on the soil surface following a rainfall event, and to control runoff without causing erosion. Surface drainage can affect the water table by reducing the volume of water entering the soil profile. This type of improvement includes: land leveling and smoothing; the construction of surface water inlets to subsurface drains; and the construction of shallow ditches and grass waterways, which empty into open ditches and streams.

Land smoothing or leveling is a water management practice designed to remove soil from high spots in a field, and/or fill low spots and depressions where water may pond. Shallow ditches may be constructed to divert excess water to grass waterways and open ditches, which often empty into existing surface water bodies.

Some disadvantages of surface drainage improvements exist. First, these improvements require annual mainte-nance and must be carefully designed to ensure that erosion is controlled. Second, extensive earthmoving activities are expensive, and land grading might expose less fertile and less productive subsoils. Further, open ditches may interfere with moving farm equipment across a field.

The objective of subsurface drainage is to drain excess water from the plant root zone of the soil profile by artificially lowering the water table level. Subsurface drainage improvement is designed to control the water table level through a series of drainage pipes (or tubing) that are installed below the soil surface, usually just below the root zone. For Ohio conditions, subsurface drainpipe is typically installed at a depth of 30 to 40 inches, and at a spacing of 20 to 80 feet. The subsurface drainage network generally outlets to an open ditch or stream. Subsurface drainage improvement requires some minor maintenance of the outlets and outlet ditches. For the same amount of treated acreage, subsurface drainage improvements generally are more expensive to construct than surface drainage improvements.

Whether the drainage improvement is surface, subsurface or a combination of both, the main objective is to remove excess water quickly and safely to reduce the potential for crop damage. In a situation where water is ponded on the soil surface immediately following a rainfall event, a general rule of thumb for most agricultural crops grown in Ohio is to lower the water table to 10 to 12 inches below the soil surface within a 24-hour period, and 12 to 18 inches below the soil surface within a 48-hour period. Properly draining excess water from the soil profile where plant roots grow helps aerate the soil and reduces the potential for damage to the roots of growing crops. Further, proper drainage will produce soil conditions more favorable for conducting farming operations. In states that depend heavily on irrigation, subsurface drainage is often used to prevent harmful buildup of salt in the soil.

Like the surface field drainage systems, the subsurface field drainage systems can also be differentiated in regular systems and checked (controlled) systems.

Mug and sole drain which was widely used in Scotland during 18th Century

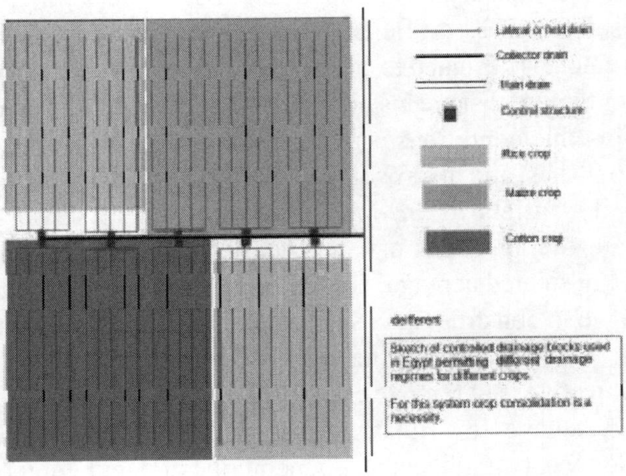

Controlled drainage system

When the drain discharge takes place entirely by gravity, both types of subsurface systems have much in common, except that the checked systems have control gates that can be opened and closed according to need. They can save much irrigation water. A checked drainage system also reduces the discharge through the main drainage system, thereby reducing construction costs.

When the discharge takes place by pumping, the drainage can be checked simply by not operating the pumps or by reducing the pumping time. In northwestern India, this practice has increased the irrigation efficiency and reduced the quantity of irrigation water needed, and has not led to any undue salinization.

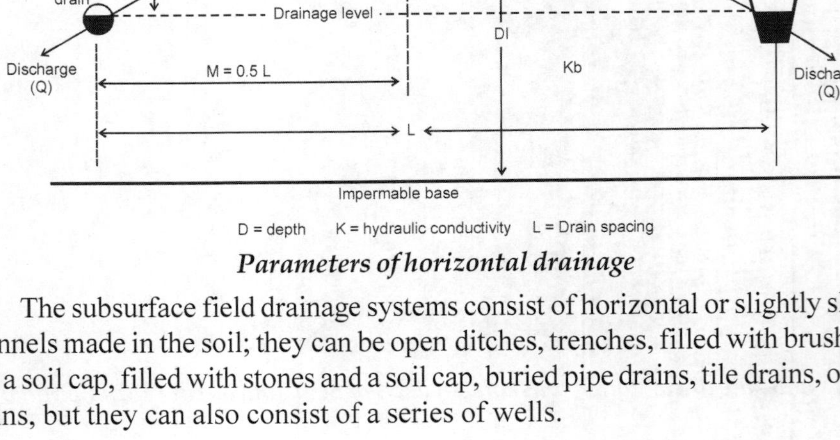

Parameters of horizontal drainage

The subsurface field drainage systems consist of horizontal or slightly sloping channels made in the soil; they can be open ditches, trenches, filled with brushwood and a soil cap, filled with stones and a soil cap, buried pipe drains, tile drains, or mole drains, but they can also consist of a series of wells.

Modern buried pipe drains often consist of corrugated, flexible, and perforated plastic (PE or PVC) pipe lines wrapped with an envelope or filter material to improve the permeability around the pipes and to prevent entry of soil particles, which is especially important in fine sandy and silty soils. The surround may consist of synthetic fibre (geo textile).

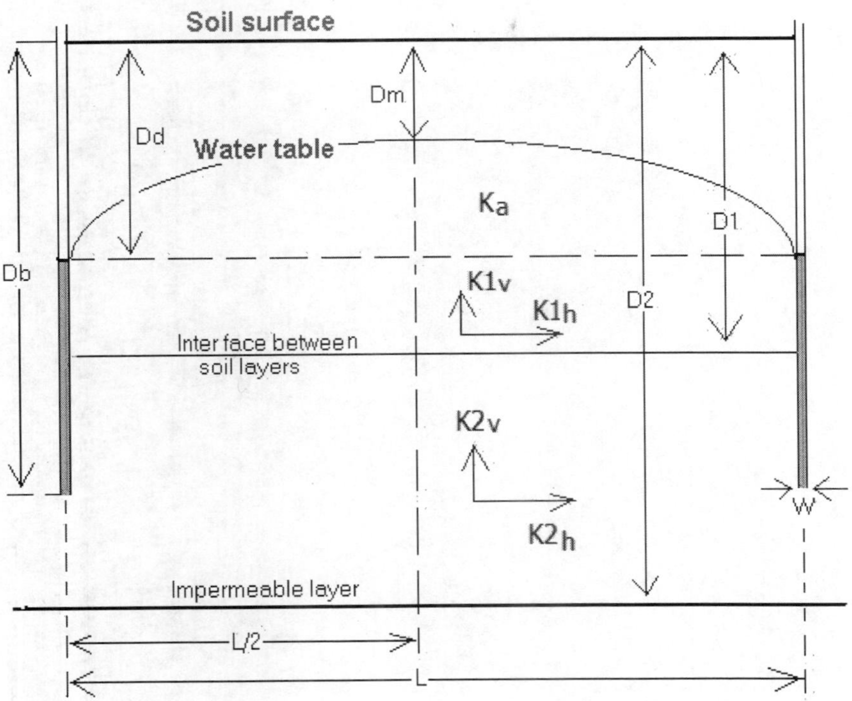

Geometry well drainage system

D = Depth K = hydraulic conductivity L = well spacing W = well diameter

Parameters of vertical drainage

The field drains (or laterals) discharge their water into the collector or main system either by gravity or by pumping.

The wells (which may be open dug wells or tube wells) have normally to be pumped, but sometimes they are connected to drains for discharge by gravity.

Subsurface drainage by wells is often referred to as vertical drainage, and drainage by channels as horizontal drainage, but it is more clear to speak of "field drainage by wells" and "field drainage by ditches or pipes" respectively.

In some instances, subsurface drainage can be achieved simply by breaking up slowly permeable soil layers by deep plowing (sub-soiling), provided that the underground has sufficient natural drainage. In other instances, a combination of sub-soiling and subsurface drains may solve the problem.

Main Drainage Systems

The main drainage systems consist of deep or shallow collectors, and main drains or disposal drains.

Deep collector drains are required for subsurface field drainage systems, whereas shallow collector drains are used for surface field drainage systems, but they can also be used for pumped subsurface systems. The deep collectors may consist of open ditches or buried pipe lines.

Deep collector drain

The terms deep collectors and shallow collectors refer rather to the depth of the water level in the collector below the soil surface than to the depth of the bottom of the collector. The bottom depth is determined both by the depth of the water level and by the required discharge capacity.

The deep collectors may either discharge their water into deep main drains which are drains that do not receive water directly from field drains, but only from collectors, or their water may be pumped into a disposal drain.

Disposal drains are main drains in which the depth of the water level below the soil surface is not bound to a minimum, and the water level may even be above the soil surface, provided that embankments are made to prevent inundation. It can serve both subsurface and surface field drainage systems.

Deep main drains can gradually become disposal drains if they are given a smaller gradient than the land slope along the drain.

The technical criteria applicable to main drainage systems depend on the hydrological situation and on the type of system.

Main Drainage Outlet

The final point of a main drainage system is the gravity outlet structure or the pumping station.

Applications

Surface drainage systems are usually applied in relatively flat lands that have soils with a low or medium infiltration capacity, or in lands with high-intensity rainfalls that exceed the normal infiltration capacity, so that frequent waterlogging occurs on the soil surface.

Subsurface drainage systems are used when the drainage problem is mainly that of shallow water tables.

When both surface and subsurface waterlogging occur, a combined surface/subsurface drainage system is required.

Sometimes, a subsurface drainage system is installed in soils with a low infiltration capacity, where a surface drainage problem may improve the soil structure and the infiltration capacity so greatly that a surface drainage system is no longer required.

Applications

Surface drainage systems are usually applied in relatively flat lands that have soils with a low o r medium infiltration capacity, or in lands with high-intensity rainfalls that exceed the normal infiltration capacity, so that frequent waterlogging occurs on the soil surface.

Subsurface drainage systems are used when the drainage problem is mainly that of shallow watertables. When both surface and subsurface waterlogging occur, a combined surface/subsurface drainage system is required. Sometimes, a subsurface drainage system installed in soils with a low infiltration capacity and a surface drainage problem improves the soil structure and the infiltration capacity so greatly that a surface drainage system is no longer required. On the other hand, it can also happen that a surface drainage system diminishes the recharge of the groundwater to such a n extent that the subsurface drainage problem is considerably reduced or even eliminated.

The choice between a subsurface drainage system by pipes and ditches or by tubewells is more a matter of technical criteria and costs than of agricultural criteria, because both types of systems can be designed to meet the same agricultural criteria and achieve the same benefits.

Usually, pipe drains or ditches are preferable to wells. However, when the soil consists of a poorly permeable top layer several metres thick, overlying a rapidly permeable and deep subsoil, wells may be a better option, because the drain spacing required for pipes or ditches would be very narrow whereas the well spacing can be very wide.

When the land needs a subsurface drainage system, but saline groundwater is present at great depth, it is better to employ a shallow, closely-spaced system of pipes or ditches instead of a deep, widely-spaced system. The reason is that the deeper systems produce a more salty effluent than the shallow systems. Environmental criteria may then prohibit the use of the deeper systems.

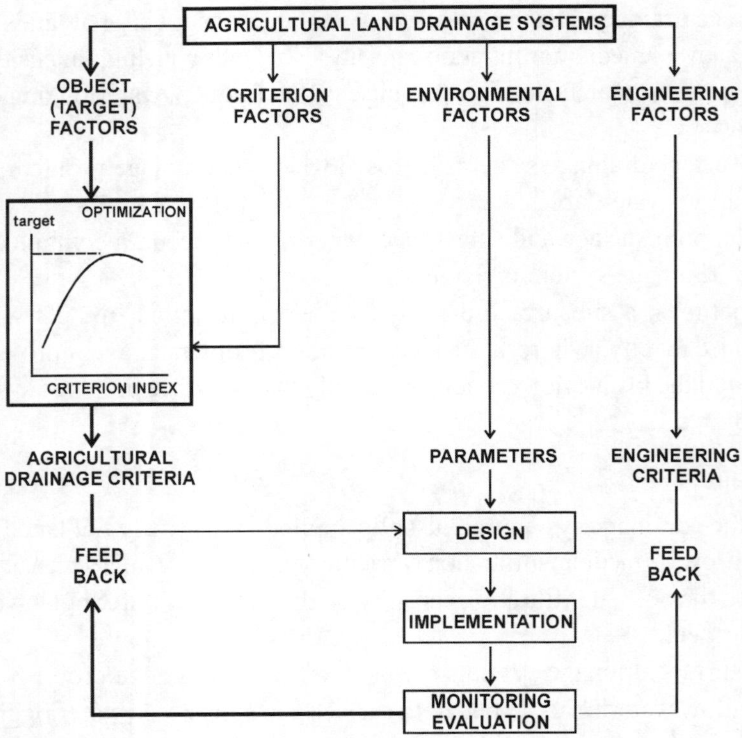

Drainage design procedures

In some drainage projects, one may find that only main drainage systems are envisaged. The agricultural and is then still likely to suffer from field drainage problems. In other cases, one may find that field drainage systems are ineffective because there is no main drainage system. Ineither of these cases, the installation of an incomplete drainage system is not recommended.

On the other hand, it can also happen that a surface drainage system diminishes the recharge of the groundwater to such an extent that the subsurface drainage problem is considerably reduced or even eliminated.

The choice between a subsurface drainage system by pipes and ditches or by tube wells is more a matter of technical criteria and costs than of agricultural criteria, because both types of systems can be designed to meet the same agricultural criteria and achieve the same benefits.

Usually, pipe drains or ditches are preferable to wells. However, when the soil consists of a poorly permeable top layer several meters thick, overlying a rapidly permeable and deep subsoil, wells may be a better option, because the drain spacing required for pipes or ditches would be considerably smaller than the spacing for wells.

Why Deeper Systems

When the land needs a subsurface drainage system, but saline groundwater is present at great depth, it is better to employ a shallow, closely spaced system of pipes or ditches instead of a deep, widely spaced system. The reason is that the deeper systems produce a more salty effluent than the shallow systems. Environmental concerns may then stop the use of the deeper systems.

In some drainage projects, one may find that only main drainage systems are envisaged. The agricultural land is then still likely to suffer from field drainage problems. In other cases, one may find that field drainage systems are ineffective because there is no adequate main drainage system. In either case, the installation of drainage systems is not recommended.

Nature of Agricultural Drainage in Ohio : A Case Study

Currently, public concern has increased about the nature of agricultural drainage, and the impact of agricultural drainage improvements on the quality of Ohio's water resources and environment.

Agriculture is Ohio's largest industry as much of the state is characterized by fertile, flat soils and adequate rainfall, crop production occurs on 45 per cent of Ohio's land area. About 55 per cent of Ohio's agricultural soils need drainage improvement to minimize soil erosion, excess soil-water conditions in the plant root zone, and unfavorable field conditions for farm equipment in the spring and fall. Ohio's agricultural drainage needs are very similar to those in states such as Arkansas, Florida, Indiana, Illinois and Louisiana. Nationally, drainage improvement is required on more than 20 per cent of US cropland (approximately 110 out of 421 million acres).

Maintaining existing water management improvements is quite important because proper management of the soil, and soil water, is required to sustain production and profitability on agricultural soils.

This publication is designed to help Ohio's citizens understand the purpose and nature of agricultural drainage improvements, particularly those related to the drainage of excess water from cropland in Ohio.

Healthier Environment for Plant Growth

In Ohio, most agricultural producers improve the drainage on their land to help create a healthier environ-ment for plant growth and to provide drier field conditions so farm equipment can access the farm field throughout the crop production season.

Healthy, productive plants have the potential to produce greater yields and more food. Also, research in Ohio has shown that agricultural drainage improvement can help reduce the year-to-year variability in crop yield, which helps reduce the risks associated with the production of abundant, high quality, affordable food. Improved access of farm equipment to the field provides more time for field activities, can help extend the crop production season, and helps reduce crop damage at harvest.

Types of Improvements

In Ohio, the two primary types of agricultural drainage improvement are surface and subsurface. Many times a land owner installs a combination of these two types.

Land drainage activities have impacted Ohio's environment and water resources. Early settlers began draining Ohio's swamps in the 1850s, and today approximately 90 per cent of Ohio's wetlands have been converted to other uses. This loss is attributed to public health considerations; rural, urban and industrial development; and agriculture.

Today, however, an important distinction needs to be made between improving the drainage of wet soils presently in agricultural production and converting our "true" remaining wetlands for other purposes. True wetlands, like bogs, marshes and swamps, have saturated soil conditions over a long enough period of time during the year to maintain water-loving vegetation and wildlife habitat. These areas, once their benefit is determined, should be protected from development.

Benefits for Environment

Wetlands provide many benefits for the environment, including wildlife habitat and enhanced water quality. An important water quality function of wetlands is the trapping and filtering of sediment, nutrients and other pollutants that enter runoff from agricultural, construction and other rural and urban sources. Interestingly, subsurface drainage improvements, in a more limited capacity, provide some of these same water quality benefits while providing a necessary element for sustained agricultural production on a majority of Ohio's productive agricultural soils.

Present agricultural trends are toward more intensive use of Ohio's existing cropland, with much of the emphasis on management. Maintaining existing agricultural drainage improvements and improving the drainage on wet agricultural soils presently in agricultural production helps minimize the need for landowners to convert additional land to agricultural production. In many cases, restoration of previously converted wetlands would be impossible because of large-scale channel improvements, urbanization and Lake Erie shoreline modification. The focus should be placed on protecting existing true wetlands and establishing new wetland areas, while maintaining our highly productive agricultural areas.

Throughout Ohio and the Midwest, the removal of excess water from wet agricultural soils is essential for providing a healthy environment for crop growth, and subsequently, helps provide affordable, high-quality food. Agricultural drainage improvement is necessary to sustain agricultural production.

Factors Contributing to Excess Water Problems in Soils

The factors that contribute to excess soil water problems includes : fine soil texture; massive soil structure; low soil permeability; topography; soil compaction; restrictive geologic layer and excess precipitation.

Soil Texture: The sand, silt and clay composition of the solid mineral particles in a soil is called soil texture. For a loam soil texture, for instance, the mineral content might consist of 40 per cent clay, 30 per cent silt and 30 per cent sand.

Soil texture can have a dramatic effect on how well the soil holds water, and how easily water can move through the soil. Fine-textured soils have a large per centage of clay and silt particles. These soils generally hold water well, but drain poorly. Coarse-textured soils have a large per centage of sand or gravel particles. These soils drain well, but have poor water-holding ability.

Soil Structure: The physical arrangement of the solid mineral particles of a soil is the soil structure. A granular structure helps promote the movement of water through a soil, but a structure that is massive (lacking any distinct arrangement of soil particles) usually decreases the movement of water.

Permeability: In general terms, the relative ease with which water can move through a block of soil is soil permeability. A soil's permeability can be affected by its texture, structure, human activities, and other factors.

Topography: The shape and slope of the land surface can cause wet soil conditions, especially around depressions where water tends to accumulate. Without an outlet, the water may drain away very slowly.

Geologic Formation: The geological formation underlying a soil can impact the drainage of water from that soil. For instance, a soil could have texture and structure properties that are beneficial to the movement of water.

However, if the geologic formation underlying this soil consisted of dense clay or solid rock, it could restrict the downward movement of water, causing the soil above the formation to remain saturated during certain times of the year.

Compaction: Human activities may help create excess soil water problems. For example, operating equipment on a wet soil can compact the soil and destroy its structure.

A soil layer that is compacted will generally have no structure, and most of the voids in this layer will have been eliminated. Voids are open spaces between soil particles that can be filled with air, water, or a combination of both.

Soil water will tend to accumulate above the compacted layer because movement of water through the compacted layer is severely restricted. If the compacted layer is located at the soil surface, very little water will enter the soil and much of the water will runoff, potentially creating a flooding and/or erosion hazard.

Precipitation: Ohio's average annual precipitation is 38 inches, based on a 50-year period of precipitation records. Even though the distribution of precipitation across the state varies, the state receives an abundant supply of precipitation in an average year.

In an average month, most areas of the state receive two to four inches of precipitation. This amount is adequate to sustain high crop yields. However, excessive rainfall, and/or winters with heavy snowfall, often produce excess soil water conditions. Furthermore, thunderstorms will frequently result in runoff because the rainfall rate is greater than the rate at which water can enter infiltrate into the soil.

BENEFITS OF AGRICULTURAL DRAINAGE

Drought Prevention

Good drainage promotes good root growth as crops reach for the capillary water in the soil. In a field with high water tables young plants only develop shallow root systems. When dry weather arrives shallow roots can not supply enough water to the crops. Consequently, it withers and dies. Crops in well drained soils have a better root system to reach the required water.

Fertilizer Efficiency Good drainage makes any fertilizer work better. Increased fertilizer usage is no alternative to good drainage—they must be used hand in hand. Healthier soil increases the effectiveness of fertilizers, and at today's consistently climbing cost it is imperative that the maximum benefit is derived from every ounce supplied.

Environmental Control Drainage reduces surface run off by increasing the water holding capacity of the soil so erosion is dramatically decreased in many cases. During heavy rain, water is removed by being absorbed into the soil and filtering down through the soil to the drain, rather than running over a saturated surface and washing away valuable soil.

Increased Quality Good drainage contributes to quality. Regardless of the field, healthy roots support healthy crops. Well drained soil discourages plant disease. Also there is a direct relationship between drainage and crop quality. Time and Labour Savings Good drainage eliminates the need to work around and then return to wet spots. Profitability can be increased through faster field work and a reduced need for repeat operations such as reseeding spots. Fertilizer Efficiency Good drainage makes any fertilizer work better. Increased fertilizer usage is no alternative to good drainage- they must be used hand in hand.

Healthier soil increases the effectiveness of fertilizers, and at today's consistently climbing cost it is imperative that the maximum benefit is derived from every ounce supplied. Environmental Control Drainage reduces surface run off by increasing the water holding capacity of the soil so erosion is dramatically decreased in many cases. During heavy rain, water is removed by being absorbed into the soil and filtering down through the soil to the drain, rather than running over a saturated surface and washing away valuable soil.

Less Soil Damage Good drainage avoids unnecessary soil damage such as compaction, smearing and crusting. Working in wet fields causes damage to soil structure which may take years to put right. Also, equipment which becomes bogged down means costly, time-consuming delays and often expensive damage to the machineryCost Effectiveness Expenditure of field drainage should be looked upon as a capital investment. It can bring benefits for many years.

Proper Functioning of Pipes

Methods are needed to not only locate buried agricultural drainage pipe, but to also determine if the pipes are functioning properly with respect to water delivery. The primary focus of this research project was to confirm the ability of ground penetrating radar (GPR) to locate buried drainage pipe, and then determine if GPR provides insight into drain line water conveyance functionality. Ground penetrating radar surveys using 250 MHz transmitter/receiver antennas were conducted at a specially designed test plot under drained, moderately wet soil conditions (water table below drain lines) and undrained, extremely wet soil conditions (water table above drain lines).

The test plot contained four drain lines: one a clay tile to corrugated plastic tubing (CPT) drain line that was completely open to flow; one comprised of CPT with an isolated obstruction near the midpoint, completely preventing through-flow of water; one comprised of CPT but filled with soil; and one comprised of CPT but severed near its midpoint, producing a partial obstruction to water flow.

Subsequent GPR computer modeling simulations were employed to assist with interpretation of the GPR field data. Results of the GPR field surveys indicate that given suitable shallow hydrologic conditions, GPR not only finds drainage pipes, but can also determine the position along a drain line where there is an isolated obstruction that completely blocks water flow. However, results show that a partial pipe obstruction is difficult to locate using GPR. Surprisingly, the soil-filled drain line was clearly detectable under both soil hydrologic conditions tested. The GPR computer modeling simulations indicate that soil had likely settled within the pipe, and that the GPR responses obtained at the test plot for the soil-filled pipe were responses representative of a pipe that was in fact only partially filled with soil. Overall, these research results provide valuable information for those contemplating the use of GPR to locate agricultural drainage pipes and then determine their functionality.

What is Agricultural Drainage Criteria?

It can be defined as criteria specifying the highest permissible levels of the water table, on o r in the soil, so that the agricultural benefits are not reduced by problems of waterlogging.

If the actual water levels are higher than specified by the criteria, an agricultural drainage system may have to be installed, or an already installed system may have to be improved, so that the water logging is eliminated. If, on the other hand, a drainage system has lowered water levels to a depth greater than specified by the criteria, we speak of a n over-designed system.

Besides employing agricultural drainage criteria, we also employ technical drainage criteria to minimize the costs of installing and operating the system, while maintaining the agricultural criteria, environmental drainage criteria (to minimize the environmental damage), and economic drainage criteria to maximize the net benefits.

Agricultural drainage systems' are systems which make it easier for water to flow from the land, so that agriculture can benefit from the subsequently reduced water levels. The systems can be made to ease the flow of water over the soil surface or through the underground, which leads to a distinction between 'surface drainage systems' and 'subsurface drainage systems'. Both types of systems need an internal or 'field drainage system', which lowers the water level in the field, and an external or 'main drainage system', which transports the water to the outlet.

Agricultural drainage systems do not necessarily lead to increased peak discharges. Although, this may occur, especially with surface drainage, the reduced waterlogging can lead to an increase in the storage of water on or in the soil during periods of peak rainfall, so that peak discharges are indeed reduced. A drainage engineer should see to it that the flow of water from the soil occurs as steadily a s possible instead of suddenly.

Sometimes like in irrigated, ponded rice fields, a form of temporary drainage is required whereby the drainage system is only allowed to function on certain occasions like during the harvest period. If allowed to function continuously, excessive quantities of water would be lost.

Such a system is therefore called a 'checked drainage system'. More usually, however, the drainage system should function as regularly as possible to prevent undue water logging at any time. We then speak of a 'regular drainage system'. In literature, this is sometimes also called 'relief drainage'.

The above definition of agricultural drainage systems excludes drainage systems for cities, highways, sports fields, and other non-agricultural purposes. Further it excludes natural drainage systems. Agricultural drainage systems are artificial and are only installed when the natural drainage is insufficient for a satisfactory form of agriculture. The definition also excludes such reclamation measures as 'hydraulic erosion control' (which aims rather at reducing the flow of water from the soil than enhancing it) and 'flood protection' (which does not enhance the flow of water from the soil, but aims rather at containing the water in watercourses). Nevertheless, flood protection and drainage systems are often simultaneous components of land reclamation projects. The reason is that installing drainage systems without flood protection in areas prone to inundation would be a waste of time and money.

Areas with both flood protection and drainage systems are often called 'polders'. Sometimes, a flood-control project alone suffices to cure the waterlogging. Drainage systems are then not required.

In literature, one encounters the term 'interceptor drainage'. The interception and diversion of surface waters with catch canals is common practice in watermanagement projects, but it is a flood-protection measure rather than a drainage measure.

The interception of ground-water flowing laterally through the soil is usually not effective, because of the low velocities of groundwater flow (seldom more than 1 m/d and often much less). In the presence of a shallow impermeable layer, subsurface interceptor drains catch very little water and generally do not relieve water logging in extensive agricultural areas.

In the presence of a deep impermeable layer, the total flow of ground water can be considerable, but then it passes almost entirely underneath the subsurface interceptor drain. The upward seepage of groundwater cannot be intercepted by a single interceptor drain: here, one needs a regular drainage system.

Use of Check Gates

The checked surface drainage systems consist of check gates placed in the bunds surrounding flat basins, such as those used for rice fields in flat lands. These fields are usually submerged and only need to be drained o n certain occasions (e.g. at harvest .time). Checked surface drainage systems are also found in terraced lands used for rice.

Not much information can be found on the relationship between the various regular surface field drainage systems, the reduction in the degree of water logging, and the agricultural or environmental effects. It is therefore difficult to develop sound agricultural criteria for the regular surface field drainage systems.

Most of the known criteria for these systems concern the efficiency of the techniques of land levelling and earthmoving. Similarly, agricultural criteria for checked surface drainage systems are not very well known.

Like the surface field drainage systems, the subsurface field drainage systems can also be differentiated in regular systems and checked systems. When the drain discharge takes place entirely by gravity, both types of subsurface systems have much in common, except that the checked systems have control gates that can be opened and closed according to need. They can save much irrigation water. A checked drainage system also reduces the discharge through the main drainage system, thereby reducing construction costs.

When the discharge takes place by pumping, the drainage can be checked simply by not operating the pumps or by reducing the pumping time. In North-West India, this practice has increased the irrigation efficiency and reduced the quantity of irrigation water needed, and has not led to any undue salinization.

The subsurface field drainage systems consist of horizontal or slightly sloping channels made in the soil; they can be open ditches, buried pipe drains, or mole drains; they can also consist of a series of wells.

The channels discharge their water into the collector or main system either by gravity or by pumping. The wells which may be open dug wells or tubewells have to be pumped, but sometimes they are connected to drains for discharge by gravity. In some instances, subsurface drainage can be achieved simply by breaking up slowly permeable soil layers by deep ploughing subsoiling, provided that the underground has sufficient natural drainage. In other instances, a combination of sub soiling and subsurface drains may solve the problem.

Subsurface drainage by wells is often referred to as 'vertical drainage', and drainage by channels as 'horizontal drainage', but it is better to speak of 'field drainage by wells', or 'field drainage by ditches or pipes'.

The main drainage systems consist of deep or shallow collectors, and main drains or disposal drains. Deep collectors are required for subsurface field drainage systems, whereas shallow collectors are used for surface field drainage systems, but they can also be used for pumped subsurface systems. The terms deep and shallow collectors refer rather to the depth of the water level in the collector below the soil surface than to the depth of the bottom of the collector. The bottom depth is determined both by the depth of the water level and by the required discharge capacity.

Disposal Drains

The deep collectors may either discharge their water into deep main drains (which are drains that do not receive water directly from field drains, but only from collectors), or their water may be pumped into a 'disposal drain'.

Disposal drains are main drains in which the depth of the water level below the soil surface is not bound to a minimum, and the water level may even be above the soil surface, provided that embankments are made to prevent inundations. Disposal drains can serve both subsurface and surface field drainage systems. Deep main drains can gradually become disposal drains if they are given a smaller gradient than the land slope along the drain. The final point of a main drainage system is the gravity outlet structure or the pumping station. The technical criteria applicable to main drainage systems depend on the hydrological situation and on the type of system.

ANALYSIS OF AGRICULTURAL DRAINAGE SYSTEMS

Objectives and Effects

The objectives of agricultural drainage systems are to reclaim and conserve land for agriculture, to increase crop yields, to permit the cultivation of more valuable crops, to allow the cultivation of more than one crop a year, and/or to reduce the costs of crop production in otherwise waterlogged land. Such objectives are met through two direct effects and a large number of indirect effects.

The direct effects of installing a drainage system in waterlogged land are : (i) A reduction in the average amount of water stored on or in the soil, inducing drier.

- A discharge of water through the system. soil conditions and reducing waterlogging.

The direct effects are mainly determined by the hydrological conditions, the hydraulic properties of the soil, and the physical characteristics of the drainage system. The direct effects trigger a series of indirect effects. These are determined by climate, soil, crop, agricultural practices, and the social, economic and environmental conditions.

Assessing the indirect effects including the extent to which the objectives are met is therefore much more difficult, but not less important, than assessing the direct effects.

The indirect effects, which can be physical, chemical, biological, and/or hydrological, can be either positive or negative. Some examples are:

- Positive effects owing to the drier soil conditions: increased aeration of the soil; stabilized soil structure; higher availability of nitrogen in the soil; higher and more diversified crop production; better workability of the land; earlier planting dates; reduction of peak discharges by an increased temporary storage of water in the soil;
- Negative effects owing to the drier soil conditions: decomposition of organic matter; soil subsidence; acidification of potential acid sulphate soils; reduced irrigation efficiency; increased risk of drought; ecological damage;
- The indirect effects of drier soil conditions on weeds, pests, and plant diseases: these can be both positive and negative; the net result depends on the ecological conditions;
- Positive effects owing to the discharge: removal of salts or other harmful substances from the soil; availability of drainage water for various purposes; and
- Negative effects owing to the discharge: excessive leaching of valuable nutrients from the soil; downstream environmental damage by salty or otherwise polluted drainage water; the presence of ditches, canals, and structures impeding accessibility and interfering with other infrastructural elements of the land.

Many of the indirect effects are mutually influenced and also exert their influence on the direct effects. For example, as a result of drainage, the following may happen:

(a) The more intensive agriculture increases the evapotranspiration and consequently may reduce the discharge, unless this leads to an increased irrigation intensity;

(b) The more stable soil structure may increase the infiltration and the subsurface drain discharge, and decrease the surface runoff.

AGRICULTURAL CRITERION FACTORS AND OBJECT FUNCTIONS

In agricultural drainage, one is dealing with agricultural, environmental, engineering, economic and social aspects.

The agricultural aspects concern ‘object factors’ and ‘criterion factors’. Object factors represent the agricultural aims that are to be achieved to the highest possible degree (maximization) through a process of optimization, yielding ‘agricultural targets’.

Optimizing is done with criterion factors, which are factors that are affected by the drainage system and at the same time influence the object factors.

Examples of criterion factors are the degree of water logging, the dryness or wetness of the soil, and the soil salinity.

Owing to its variation in time and space, a criterion factor can be specified in different ways. A chosen specification can be called a criterion index'. Examples of such indices are:

- The average depth of the watertable during the cropping season;
- The average depth of the watertable during the off-season;
- The exceedance frequency of the watertable over a critically high level;
- Seasonal average salinity of the rootzone;
- Salinity of the topsoil at sowing time; and
- Average, minimum, or maximum number of days that the soil is workable during a critical period.

The relationship between a n object factor and an index can be called 'object function of the index' and is also known as 'response function' or 'production function'.

The optimization procedure through the object function leads to a tolerance, or even an optimum, value of the index, which can be called as an 'agricultural drainage criterion'. It serves as an instruction to the designer of the drainage system because it stipulates the agricultural condition the system must meet to be effective (i.e. to fulfil its purpose). Also, the instruction can prevent the design and implementation of a system that is unnecessarily intensive, expensive and even detrimental.

'Environmental factors' are factors representing the given natural or hydrological conditions under which the system has to function. Examples of these factors are irrigation, rainfall, the soil's hydraulic conductivity, natural surface or subsurface drainage, topography, and aquifer conditions. For design purposes, the environmental factors must be specified as 'environmental indices', in the same way as the criterion factors are specified as criterion indices.

Examples of environmental indices are the average seasonal rainfall, the extreme daily rainfall, the arithmetic or geometric mean of the hydraulic conductivity, and the variation in hydraulic conductivity with depth in the soil. Through a process of optimizing the engineering aspects, the environmental indices yield 'environmental parameters', which are fixed values of the indices, chosen as engineering or design criteria, in similarity to the agricultural criteria. Examples of such parameters are design values for rainfall, discharge, and hydraulic conductivity.

The engineering aspects include 'engineering factors' and 'engineering objectives'. The objectives usually aim at minimizing the costs, and relate to the efficiency of the drainage system. A fully efficient drainage system fulfils the agricultural criteria at the lowest possible input level of materials and finances.

The engineering factors are factors representing the technical and material components of the drainage system like the layout, the longitudinal section and the cross-section of the drains, and the kind and quality of materials. The choice of the engineering factors is specified in the tender documents produced after the design has been completed.

NATURAL DRAINAGE SYSTEM

Drainage is the process of draining the water from soaked land, through pipes, tunnels, canals, ditches and trenches. The patterns formed by streams, rivers and lakes are called as natural drainage systems. Implementation of proper drainage systems can prevent the municipal plus the rural areas from flooding. Geosynthtics are frequently used to improve performance and extend the life of drainage systems. Drainage or subsurface drainage, agricultural hydraulics, is the combination of natural and artificial systems that makes possible deep disposal of excess land. A well-drained soil occurs naturally when the water table is at proper depth and the porosity is such as which will not slow down the percolation of excess water.

There are various types of drainage systems present for natural resources. Dendritic drainage system is the most common form of drainage system in which, many contributing streams joined together into the tributaries to form larger rivers. Dendritic system forms V- shaped valley. As a result of that the rock will become non porous. Parallel drainage system is the type of drainage system in which the pattern caused by rivers resulting in steep slopes. Because of the steep slopes, all the streams flow in the same direction. The streams of this type are swift, straight and consist of very few tributaries.

Trellis is a type of drainage system, which is similar to that of a common garden trellis. When the river flows along a valley, smaller tributaries feeds into it from the steep slopes on the sides of the mountains. These tributaries join the river at about 90 degrees angle. This appears like the drainage system and is the characteristic of the folded mountains pattern. Rectangular drainages develop on rocks, which are approximately uniform resistance to erosion but usually have two directions of joining at right angles.

Gradually, the streams develop along these joints and are comparatively less resistant to erosion than the bulk rock. Further this results in, joining of the streams with riverbed at right angle. The streams radiate is outwards from a central high point in the radial drainage system. The structure of the radial drainage system is like volcanoes. The radial drainage may lead a combination of radial and annular patterns. In Deranged drainage system, there is no coherent pattern to the rivers and lakes. This type of drainage systems are found in the area where there has been much geological disruption occurred.

Domestic drainage systems designed to carry wastewater from a land to the main sewage network. Depending upon the age of the property, domestic drainage systems are classed in two different types, a single-stack and a Two-pipe system. A Two pipe drainage systems were found in the houses built before late 1950s. It consists of the large diameter vertical soil pipe, which leads directly to the sewage network, and often runs down the outside wall of a property of all any new houses that are built with a single-stack drainage system. Unlike the two-pipe system, a large bore pipe often incorporated into the internal framework of a property. For cleaning the drainage systems the chemical unblockers are the good options but pumps or plungers can be used to clean the drainage systems.

Importance of Rainwater Drainage Systems

Property is always a great investment and generally it is quite a large one, if not your largest. If you own your own property you know how important it is to maintain it as well as you possibly can. So you must take care of it to ensure that it does not become damaged or badly kept. If you allow this to happen the value of your investment can be seriously affected and this is something that you must avoid.

One of your worst enemies as a property owner is the weather. Adverse weather conditions can cause a great deal of damage to your property if you do not have the proper equipment in place to protect it. Heavy rain is one of the most dangerous types of weather as moisture within your building can cause as huge amount of damage. Because of this having a high quality drainage system in place is absolutely paramount. These drainage systems are designed and manufactured to provide your building with the maximum amount of protection possible from heavy or constant rain fall.

As there are son many different types of drainage systems available on the current market place it can be a very difficult decision trying to pick the best one, especially if you are relatively new to the property game. There are a number of different factors that you will need to take into consideration when you are choosing your rainwater drainage system. Firstly, you will need to look at the style of building that you have.

It is absolutely important that the drainage system that you choose matches the style of your building. If it does not it can detract from the aesthetics of your building and thus affect the appeal and even the price of your home. This cannot be allowed to occur, so you must really take your time while browsing through the various drainage systems that are available to you.

There are plenty to choose from including modern and contemporary styled drainage systems and more traditional and classically styled drainage systems. This means that you should have no problem locating a system to suit both the style of you home and your own personal taste.

There are many different retailers from where you can get high quality rainwater drainage systems. You can purchase them from a number of retailers and home improvement stores on the high street and the selection within these stores is quite good.

However, the very best location to find not only the best selection but the best prices is the internet. Here you can gain access to the entire marketplace and you can also find sites and blogs that will offer you help and assistance that will indicate what the very best quality drainage systems are.

CHEST DRAINAGE SYSTEMS

Traditional Chest Drainage

In 1967, Deknatel introduced the first integrate ddisposable chest drainage unit based on the three bottle system. Now that we have reviewed normal anatomy, physiology, and pathophysiology, let's discuss each of the three chambers in detail.

Collection Chamber

At the right side of the unit is the collection chamber. The patient tubing connects thedrainage unit directly to the chest tube. Any drainagefrom the chest flows into this chamber. The collection chamber is calibrated and has a write-on surface toallow for easy measurement and recording of the time, date, and amount of drainage.

Water Seal Chamber

The middle chamber of a traditional chest drainage system is the water seal. The main purpose of thewater seal is to allow air to exit from the pleuralspace on exhalation and prevent air from enteringthe pleural cavity or mediastinum on inhalation.When the water seal chamber is filled with sterile fluidup to the 2 cm line, a 2 cm water seal is established.

To maintain an effective seal, it isimportant to keep the chest drainage unit upright atall times and to monitor the water level in the waterseal to check for evaporation. Bubbling in the water seal chamber indicates an airleak. The patient air leak meter indicates the approximate degree of air leak from the chest cavity. The meter is made up of numbered columns, labeled from 1 (low) to 7 (high). The higher the numbered column through which bubbling occurs, the greater the degree of air leak. Bydocumenting the number, the clinician can monitorair leak increase or decrease. The water seal chamber also has a calibrated manometer to measure the amount of negativepressure within the pleural cavity.

Thewater level in the small arm of the water seal rises asintrapleural pressure becomes more negative. If there is no air leak, the water level should rise andfall with the patient's respirations, reflecting normal pressure changes in the pleural cavity. During spontaneous respirations, the water level should riseduring inhalation and fall during exhalation. If the patient is receiving positive pressure ventilation, the oscillation will be just the opposite — the water level should fall with inhalation and rise with exhalation. This oscillation is called tidaling and is one indicatorof a patent pleural chest tube.At the top of the water seal chamber is a highnegativity float valve and high negativity relief chamber. These safety features maintain the water seal in the event of high negative pressures. Three situations can cause high negative pressure:

(a) The patient in respiratory distress, coughing vigorously, or crying; (b) Chest tube stripping; (c) Decreasing or disconnecting suction. High negativity is indicated by rising water in the small arm of the water seal chamber. If the water rises beyond

–20 cm, the high negativity float valve will rise and impede the flow of water, allowing the patient to develop as much negativity as needed for inspiration. In instances of falsely imposed high negative pressure, such as stripping chest tubes,water will continue to rise, filling the high negativity relief chamber.

The relief chamber automatically vents excessive negative pressure, thus preventing respiratory compromise from accumulated negativity.Vigorous milking or stripping can create dangerously high negative pressures. Research has documented negative pressures as high as –450 cm H2O. Pleurevac prevents accumulation of excessive high negative pressure as discussed above; however, the transient high negative pressures created by vigorous stripping can put the patient at risk for media stinal trauma and graft trauma.

Use extreme caution and follow your hospital policy. A manual high negativity relief valve is located on top of chest drainage systems. Depressing the high negativity relief valve allows filtered air into the system, relieving negativity and allowing the waterlevel to return to baseline in the water seal. Use the high negativity relief valve with caution. If suction is not operative, or if operating on gravity drainage, depressing the high negativity relief valve can reduce negative pressure within the collection chamber to zero (atmosphere) with the resulting possibility of apneumothorax.

Wet Suction Control

The chamber on the left side of the unit is the suction control chamber. Traditional chest drainage units regulate the amount of suction by the height of a column of water in the suction control chamber. It is the height of water, not the setting of the suction source, that actually limits the amount of suction transmitted to the pleural cavity. Asuction pressure of –20cm H2O is commonly recommended. Lower levels may be indicated forinfants and for patients with friable lung tissue, or if ordered by the physician. In a wet suction control system such as the Pleur-evac®A-7000/A-8000 series, fill the suction control chamber to the desired height with sterile fluid. Connect the short suction tubing to a suction source, and adjust the source suction to produce gentle bubbling in the suction control chamber. Increasing suction at the suction source will increase air flow through the system, but will have minimal effect on the amount of suction imposed on the chest cavity. Excessive source suction not only causes loud bubbling (which can disturb patients and caregivers), but also hastens evaporation of water from the suction control chamber. This results in a lower amount of suction applied to the patient as the level of water decreases. Self-sealing diaphragms are provided to adjust the water level in this chamber.

Dry Suction

The next step in the evolution of chest drainage units was the development of dry suction control chambers. Dry suction control systems provide many advantages: higher suction pressure levels can be achieved, set-up is easy, no continuous bubbling provides for quiet operation, and there is no fluid to evaporate which would decrease the amount of suction applied to the patient. Instead of regulating the level of suction with acolumn of water, the dry suction units are controlled by a self-compensating regulator. A dial to set the suction control setting is located on the upper leftside of each unit. To set the suction setting, rotate the dial until the red stripe appears in the semi-circular window at the prescribed suction level and clicks into place. Suction can be set at –10, –15,–20, –30, or –40 cm of water. The unit is pre-set at–20 cm of water when opened.Connect the short suction tubing or suction port tothe suction source. Source suction must be capableof delivering a minimum of 16 liters per minute (LPM) air flow. Increase suction source until the orange float appears in the suction control indicator window.

The unique design of the Pleur-evac dry suction control immediately responds to changes in patient pressure (patient air leak) or changes in suction pressure (surge/ decrease at the suction source). The setting of the suction control dial determines the approximate amount of suction imposed regardless of the amountof source suction — as long as the orange float appears in the indicator window. Patient situations that may require higher suction pressures of –30 or –40 cm H2O include: a large airleak from the lung surface, empyema or viscous pleural effusion, a reduction in pulmonary compliance, or anticipated difficulty in expansion of the pulmonary tissue to fill the hemithorax. In the presence of a large air leak, air flow through the Pleur-evac may be increased by increasing source suction, without increasing imposed negativity. It is not necessary to change the suction setting on the Pleur-evac unit to accommodate high air flows.

The suction control level can be changed at any time as prescribed by simply rotating the dial to the new suction setting. Confirm that the orange float remains in the suction control indicator window at the new suction setting. If suction setting is changed from a higher to a lower level, the patient negativity may remain at the higher level unless the negativity is relieved. Use the manual high negativity relief valve to reduce negativity to desired level. Both the wet suction and dry suction series of Pleurevachave a positive pressure relief valve that opens with increases in positive pressure, preventing pressure accumulation. Normally, air exits through the suction port. Obstruction of this route (i.e. a bed wheel rolls on top of the suction tube, orthe suction port is capped after suction discontinued) could cause accumulation of air in the system leading to tension pneumothorax. This safety feature allows venting of the positive pressure automatically, thus minimizing the risk of tension pneumothorax.

One-way Valve

No water is required to establish the one-way seal. Just connect the patient tube to the patient's thoracic catheter and the patient seal is established for patient protection. The one-way valve maintains the patient seal even if the unit is tipped over. Unlike a water seal system in which the seal may be lost when the unit is tipped, the Sahara dry seal protects the patient from atmospheric air. If air leak diagnostics are desired, the patient air leak meter must be filled to the "Fill" ine. The fluid in the patient air leak meter is used for air leak detection as described earlier and is not a water seal.

In the Pleur-evac Sahara, negative pressure exists in the collection chamber when the yes can be seen in the indicator indow. During gravity drainage before normal negative pressure has been re-established in the pleural cavity, the indicator may intermittently indicate negative pressure with patient respiration. During suction drainage, the pressure indicator should indicate a negative pressure continuously. The negative pressure indicator does not confirm drainage tube patency. Routinely check the drainage tube patency. The Pleur-evac Sahara system also has an automatic high negative pressure relief valve to limit the negative pressure to approximately – 50 cm of H2O. Amanual high negativity relief valve is also provided tovent excessive negativity as described earlier.

Gravity Drainage

Not all patients require suction. Suction may be discontinued to transport a patient; it may also be discontinued 24 hours before chest tube removal. Consult hospital policy to determine if an order is needed to institute or discontinue suction. If suction is discontinued, the suction tube or port should remain uncapped and free of obstructions to allow air to exit and minimize the possibility of tension pneumothorax.

DESIGN OF DRAINAGE SYSTEM

Drainage System

Drainage, it is the artificial or natural removal of surface and sub-surface water from an area. The agricultural soil requires the drainage for improving production and to manage the water supplies. The design of drainage system refers to geomorphology drainage system, agriculture drainage system and sustainable urban drainage system.

Geomorphology

The drainage system in the geomorphology is the pattern formed by the streams, rivers, and lakes in the particular drainage basin. The particular region is dominated by soft or hard rocks and the gradient of the land are governed by the topography of the land. In geomorphology, a drainage basin is the topographic region from there streams receives runoff, through flow, and ground water flow. The drainage basin are divided by topographic barrier from each other is called as watershed.

The watershed represents all of the stream branches that flow to some location along the stream channel. The drainage basin varies according to the size, shape and number of the drainage basin found in the area. This included in the design of drainage system.

Agricultural Drainage System

It is by which the water level or in the soil is controlled to increase the agricultural growth and yield crop production. There are three classification of agricultural drainage systems are available they are surface drainage system, subsurface drainage system, and main drainage system. The functioning of the surface drainage systems will start if there is an excess of rainfall or irrigation, operate entirely by gravity.

This consists of reshaped land surface and can be divided into Bedded systems, used in flat lands for crops other tan rice; graded systems, used in sloping the land for crops than rice. The subsurface drainage systems consist of drainage systems, it consist slightly sloping channels made in the soil; they can open ditches, trenches, filled with stones and soil cap, mole drains. The main drainage systems consist of deep or shallow collectors and main drains or disposal drains.

Sustainable Drainage Systems

The design of drainage system includes sustainable drainage system. This is sometime called as Sustainable Urban drainage systems; this is designed to reduce the potential impact of the new and existing developments in respect with surface water drainage discharges. To reproduce the natural systems is the idea behind sustainable drainage systems. This uses the effective cost with low environmental impact to drain away dirty and surface water collections, and cleaning them before allowing it released slowly back into environment, i.e. into water courses.

2

Water Resources and Irrigation in India

Water is indispensable for human, animals and plant life. It is a part of all organisms, some of which contain more than 90 percent. Water is essential part of protoplasm. It is an important ingredient in photosynthesis. About 400 to 500 liters of water is necessary for production of a one kilogram of plant dry matter. Water is also required for translocation of nutrient and dissipation of heat.

WATER RESOURCES OF INDIA

The average rainfall of India is 1194 mm. When considered over geographical area of 328 million hectares, this rainfall amounts to 392 million hectare meters (m. ha. m). This may round off to 400 (m. ha. m) by including the contribution of snowfall which is not yet fully determined. Out of 400 (m. ha. m) of rainfall, 75% is received during South-West Monsoon period (June to September) and rest in remaining months as shown below. A Major portion of water (215 m. ha. m) soaks into the soil, while 70 (m. ha. m) is lost as an evapo-transpiration.

Flow chart showing the water resources of India

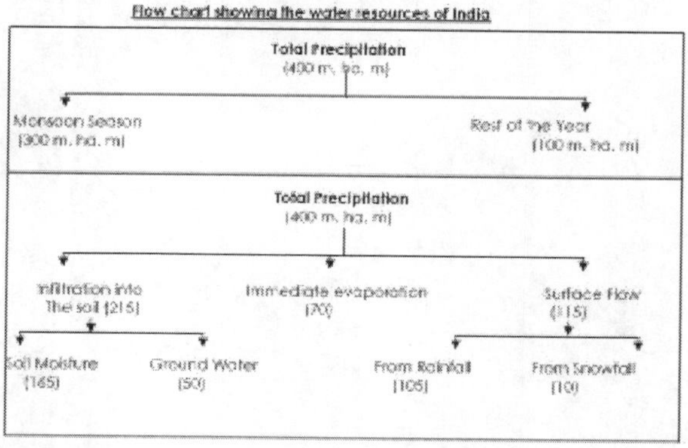

The two main sources of water in India are rainfall and glacial snowmelt in the Himalayas. Although snow and glaciers are poor producers of freshwater, they are good distributors as they yield at the time of need, in the hot season. Indeed, about 80 percent of the river flow occurs during the four to five months of the southwest monsoon season.

Several important river systems originate in upstream countries and then flow to other countries: the Indus river originates in China and flows to Pakistan; the Ganges-Brahmaputra river system originates partly in China, Nepal and Bhutan, and flows to Bangladesh; some minor rivers drain into Myanmar and Bangladesh. However, no official records are available regarding the annual flows into or our of the country.

The rivers of India can be classified into four groups:

- The Himalayan rivers (Ganges, Brahmaputra, Indus) are formed by melting snow and glaciers as well as rainfall and, therefore, have a continuous flow throughout the year. As these regions receive very heavy rainfall during the monsoon period, the rivers swell and cause frequent floods.
- The rivers of the Deccan plateau (with larger rivers such as Mahanadi, Godavari, Krishna, Pennar and Cauvery draining into the bay of Bengal in the east, and Narmadi and Tapi draining into the Arabian sea in the west), making up most of the southern-central part of the country, are rainfed and fluctuate in volume, many of them being non-perennial.
- The coastal rivers, especially on the west coast, south of the Tapi, are short with limited catchment areas, most of them being non-perennial.
- The rivers of the inland drainage basin in western Rajasthan in the northwestern part of the country, towards the border with Pakistan, are ephemeral and drain towards the salt lakes such as the Sambhar, or are lost in the sands.

For planning purposes, the country is divided into 20 river units, 14 of which are major river basins, while the remaining 99 river basins have been grouped into six river units, as presented in Table 3. The spatial imbalance of water resources distribution can be appreciated by the fact that the Ganges-Brahmaputra-Meghna basin, which covers 34 percent of the country's area, contributes about 59 percent of the water resources.

The west flowing rivers towards the Indus cover 10 percent of the area and contribute 4 percent of the water resources. The remaining 56 percent of the area contributes 37 percent to the runoff.

The potential surface water resources is assessed as the natural runoff of the rivers. Looking at the Indus Water Treaty (1960) between India and Pakistan, however, these are an estimated 1 869.37 km, of which only 690.31 km are considered usable or exploitable because of constraints related to topography, uneven distribution of the resource over space and time, geological factors and contemporary technological knowledge (Table 3).

TABLE 3
Basin wise distribution of utilisable surface water resources (Source: Central Water Commission (1993), p 12, and Central Water Commission (1996), p 15)

River basin unit	Location	Draining into	Catchment area (% of the country)	Average annual runoff (km³)	Exploitable surface water (km³)
1 Ganges	Northeast	Bangladesh	26.5	525.02	250.0
Brahmaputra	Northeast	Bangladesh	6.0	537.24	24.0
Meghna/Barak	Northeast	Bangladesh	1.5	48.36	
2 Minor rivers of the northeast	Extreme northeast	Myanmar	1.1	20.00	
		Bangladesh		11.00	
3 Subarnarekha	Northeast	Bay of Bengal	0.9	12.37	6.8
4 Brahmani-Baitarani	Northeast	Bay of Bengal	1.6	28.48	18.3
5 Mahanadi	Central-east	Bay of Bengal	4.4	66.88	50.0
6 Godavari	Central	Bay of Bengal	9.7	110.54	76.3
7 Krishna	Central	Bay of Bengal	8.0	78.12	58.0
8 Pennar	Southeast	Bay of Bengal	1.7	6.32	6.9
9 Cauvery (1)	South	Bay of Bengal	2.5	21.36	19.0
10 East flowing rivers between Mahanadi and Pennar	Central-east	Bay of Bengal	2.7	22.52	13.1
11 East flowing rivers between Kanyakumari and Pennar	Southeast	Bay of Bengal	3.1	16.46	16.7
12 West flowing rivers from Tadri to Kanyakumari	Southwest	Arabian Sea	1.7	113.53	24.3
13 West flowing rivers from Tapi to Tadri	Central-west	Arabian Sea	1.7	87.41	11.9
14 Tapi	Central-west	Arabian Sea	2.0	14.88	14.5
15 Narmada (2)	Central-west	Arabian Sea	3.1	45.64	34.5
16 Mahi	Northwest	Arabian Sea	1.1	11.02	3.1
17 Sabarmati	Northwest	Arabian Sea	0.7	3.81	1.9
18 West flowing rivers of kutch and Saurashtra	Northwest	Arabian Sea	10.0	15.10	15.0
19 Rajasthan inland basin	Northeast	-	0.0	Negligible	-
20 Indus Eastern tributaries (3a)	Northwest	Pakistan	10.0	11.10	46.0
Indus Western tributaries (3b)				62.21	
Total considering Indus Treaty			**100.0**	**1 869.37**	**690.3**

Notes: (1) The assessment for Cauvery was made by the Cauvery Fact Finding Committee in 1972 based on 38 years' flow data at Lowe Anicut on Coleroon. An area of 8000 km² in the delta is not accounted for in this assessment. (2) The potential of the Narmada basin was determined on the basis of catchment area proportion from the potential assessed at Garudeshwar as given in the report on Narmada Water disputes Tribunal Decision (1978). (3) Under the Indus Water Treaty (1960) between India and Pakistan the following is foreseen. (a) All waters of the eastern tributaries of the Indus originating in India (11.1 km³) shall be available for unrestricted use by India (Sutlej, Beas, Ravi), except some domestic and non-consumptive use. (b) All waters of the western tributaries (Chenab, Jhelum) shall be available for unrestricted use by Pakistan, except for some domestic use, non-consumptive use, agricultural use, generation of hydroelectric power; total flow of western rivers is estimated at around 232.48 km³ (flow from China to India is estimated at 181.62 km³ and flow generated within India at 50.86 km³), of which 170.27 km³ should then be reserved for Pakistan and therefore 62.21 is available for India.

Annual renewable groundwater resources are an estimated 432 km, of which around 90 percent or 390 km are considered overlap between surface water and groundwater. Annual internal renewable surface water resources (IRSWR) have been estimated as 1 446.42 km, of which 1 404.42 km surface water, 432 kmgroundwater and 390 km overlap. The IRSWR have been estimated by deducting the inflow from the total renewable surface water resources.

Under the Indus Water Treaty (1960) between India and Pakistan, it was estimated that 73.31 km/year is available for India (Table 3). The following rules apply:

- Eastern rivers: All the waters of the eastern tributaries of the Indus river originating in India, i.e. the Sutlej, Beas and Ravi rivers taken together, shall be available for unrestricted use by India. Pakistan shall be under an obligation to let flow, and shall not permit any interference with, the waters (while flowing in Pakistan) of any tributary which, in its natural course, joins the Sutlej Main or Ravi Main before these rivers have finally crossed into Pakistan. This average annual flow in India before crossing the border is an estimated 11.1 km3. All the waters, while flowing in Pakistan, of any tributary which in its natural course joins the Sutlej main or the Ravi main, after these rivers have crossed into Pakistan, shall be available for the unrestricted use of Pakistan.
- Western rivers: Pakistan shall receive for unrestricted use all those waters of the western rivers, i.e. Chenab and Jhelum, which India is under obligation to let flow, except for restricted uses, related to domestic use, non-consumptive use, agricultural use and generation of hydroelectric power of which the amounts are set out in the Treaty. Annual flow from China to India in the Indus basin is 181.62 km and it is estimated that the flow generated within India is 50.86 km, resulting in a flow from India to Pakistan in this part of 232.48 km, of which 170.27 km reserved for Pakistan and 62.21 km available for India.

Besides this outflow to Pakistan from the Indus basin, 1 121.62 km flows annually to Bangladesh (525.02 km from the Ganges, 537.24 km from the Brahmaputra, 48.36 km from the Meghna and 11 km from other rivers into southeast Bangladesh), and 20 km flows to Myanmar.

In 1996, produced wastewater was an estimated 25.4 km. In 2004, wastewater production in urban centres (rural areas with larger population have not been accounted) was an estimated 10.585 km and the treated wastewater was about 2.555 km.

No reliable statistics are available on the number or the status of desalination plants, or on their capacities or technologies adopted. Estimates indicate, however, that there are more than 1 000 membrane-based desalination plants of various capacities ranging from 20 m/day to 10 000 m/day. There are few thermal-based desalination plants. In 1996, some 550 000 m of seawater were desalinated in the Lakshadweep Islands, mainly using electro dialysis and reverse osmosis (RO). Solar stills are also installed on the peninsula, as in Gujarat in the northwest. A 6 300 m/day desalination plant is being set up at Kalpakkam, Chennai with a capacity of 4 500 m from multi stage flash (MSF) method, using low pressure steam from the Madras Atomic Power Station and 1 800 m/day from RO method. While the Plant using the RO method is under operation, the MSF-based plant is to be commissioned soon.

The total constructed water storage capacity, up to 2005, was 224 km. Another 76.26 km are estimated to be possible from dams under construction and 107.54 from dams under consideration. Seven dams have a reservoir capacity that exceeds 8 km. They are the Nagarjuna Sagar dam on the Krishna river (11.56 km), the Rihand dam on the Rihand river (10.6 km), the Bhakra dam on the Sutlej river (9.62

km), the Srisailam dam on the Krishna river (8.72 km), the Hirakud dam on the Mahanadi river (8.1 km), the Pong (Beas) dam on the Beas river (8.57 km) and the Ukai dam on the Tapti river (8.5 km).

India controls the flow of the River Ganges using a dam completed in 1974 at Farraka, 18 km from the border with Bangladesh. The Farakka barrage is a low diversion structure and not classified as a large dam.

India is endowed with rich hydropower potential, ranking fifth in the world. The gross hydropower potential was an estimated 148 700 MW as installed capacity, or 84 000 MW at 60 percent power load factor, of which the Brahamaputra, Ganges and Indus basins contribute about 80 percent. Further, small, mini and micro hydropower schemes (with a capacity of less than 3 MW) have been assessed as having almost 6 782 MW of installed capacity.

India experiences an average precipitation of 1,170 millimetres (46 in) per year, or about 4,000 cubic kilometres (960 cu mi) of rains annually or about 1,720 cubic metres (61,000 cu ft) of fresh water per person every year. Some 80 percent of its area experiences rains of 750 millimetres (30 in) or more a year. However, this rain is not uniform in time or geography. Most of the rains occur during its monsoon seasons (June to September), with the north east and north receiving far more rains than India's west and south. Other than rains, the melting of snow over the Himalayas after winter season feeds the northern rivers to varying degrees. The southern rivers, however experience more flow variability over the year. For the Himalayan basin, this leads to flooding in some months and water scarcity in others. Despite extensive river system, safe clean drinking water as well as irrigation water supplies for sustainable agriculture are in shortage across India, in part because it has, as yet, harnessed a small fraction of its available and recoverable surface water resource. India harnessed 761 cubic kilometres (183 cu mi) (20 percent) of its water resources in 2010, part of which came from unsustainable use of groundwater. Of the water it withdrew from its rivers and groundwater wells, India dedicated about 688 cubic kilometres (165 cu mi) to irrigation, 56 cubic kilometres (13 cu mi) to municipal and drinking water applications and 17 cubic kilometres (4.1 cu mi) to industry.

Vast area of India is under tropical climate which is conducive throughout the year for agriculture due to favourable warm and sunny conditions provided perennial water supply is available to cater to the high rate of evapotranspiration from the cultivated land. Though the overall water resources are adequate to meet all the requirements of the country, the water supply gaps due to temporal and spatial distribution of water resources are to be bridged by interlinking the rivers of India. The total water resources going waste to the sea are nearly 1200 billion cubic meters after sparing moderate environmental/salt export water requirements of all rivers.Food security in India is possible by achieving water security first which in turn is possible with energy security to supply the electricity for the required water pumping as part of its rivers interlinking.

Instead of opting for centralised mega water transfer projects which would take long time to give results, it would be cheaper alternative to deploy extensively shade nets over the cultivated lands for using the locally available water sources efficiently to crops throughout the year. Plants need less than 2% of total water for metabolism requirements and rest 98% is for cooling purpose through transpiration. Shade nets or polytunnels installed over the agriculture lands suitable for all weather conditions would reduce the potential evaporation drastically by reflecting the excessive and harmful sun light without falling on the cropped area.

Issues and Challenges

Water scarcity and conflicts. Water resources in India are ridden with competition and conflict at all levels - between states (water is a "state subject" in the Constitution), between sectors (industries versus farmers; cities versus rural areas), and within irrigation command areas. On a national basis, total demand for water resources is expected to exceed the utilizable potential by 2050. Groundwater is being depleted at an alarming rate. A nationwide assessment (2004) found 29 percent of groundwater blocks to be in the semi-critical, critical or overexploited categories, with the situation rapidly deteriorating. At the same time, diversion and contamination of surface water threatens the health of rivers and the people dependent upon them. Policies of data secrecy inhibit the effectiveness of agencies responsible for planning, resource allocation and disaster preparedness and add to the atmosphere of mistrust. In the absence of a credible system for defining and enforcing resource allocations, conflict will continue to arise as a result of scarcity and competition for use.

Inadequate service delivery and weak management institutions and policies. Supply augmentation continues to be the preferred response to scarcity, rather than improved management. A decades-long push to expand the country's irrigated area has left Irrigation Departments in most states focused primarily on the construction of physical works. The same focus on supply augmentation can be seen in most of the state agencies responsible for urban and rural water supply and sanitation. Delhi, for example, with 220 liters per capita per day water resource availability and at least 40% system losses, is procuring more water from new dams in the Himalayas and reservoirs in neighboring states, while giving little attention to improved management. Although significant progress was made on increasing access by rural communities to water supply over the past decade, many systems are no longer operating properly due to poor maintenance, contamination or depletion of water source. Officials acknowledge that the traditional approach of "build-neglect-rebuild" is unsustainable, inefficient and largely responsible for the poor performance of an estimated $500 billion dollars worth of assets in water resources and irrigation infrastructure. Poorly targeted subsidies continue to be the norm. Political will is required to hold service providers accountable for efficiency and delivery of good service and allow tariffs and user charges sufficient to enable proper maintenance and renewal of facilities and incentivize efficiency on the part of users.

Limited capacity to deal with new developments and pressures such as population growth, economic development, climate change. Many of the state agencies responsible for irrigation and water supply are overstaffed, but also under-qualified in certain critical areas as a result of hiring freezes. For example, there is little capacity for social and environmental management, economics, and modeling and decision sciences, as well as inadequate knowledge of the current best practices in water management. Involvement of stakeholders most affected by poor water management (e.g., water user associations, village water committees) is still the exception rather than the rule. Similarly, the capacity available in the private sector has been barely tapped as a means of improving accountability and efficiency.

Government Strategy

The Government's National Water Mission under the National Action Plan on Climate Change addresses many of these issues. With its main objective of "conservation of water, minimizing wastage and ensuring its more equitable distribution both across and within States through integrated water resources development and management," it has the potential to shift focus from supply augmentation to efficient utilization of available resources. The Mission identifies the following five goals:

(i) Comprehensive water data base in public domain and assessment of impact of climate change on water resource;
(ii) Promotion of citizen and state action for water conservation, augmentation and preservation;
(iii) Focused attention to vulnerable areas including over-exploited areas;
(iv) Increasing water use efficiency by 20%, and
(v) Promotion of basin level integrated water resources management.

The Planning Commission in its new Framework for the Twelfth Five Year Plan is likely to recommend measures that reach even further in improving oversight and regulation of the sector, increasing transparency and stakeholder involvement and targeting subsidies more effectively. Specific measures under consideration include establishment of a National Water Commission to monitor compliance with conditionality attached to central funding, systematic mapping of aquifers and disclosure of this information to farmers and other water users, establishment of Aquifer Management Associations, separation of electricity feeders for irrigation and other uses, and targets for reuse water.

As encouraging as these pronouncements and plans are, performance in the water sector in India has typically lagged behind policy pronouncements and targets. Perhaps more significant, then are the efforts underway at the national level and in some states to:

- Improve the capacity of institutions responsible for monitoring and analyzing data on groundwater resources
- Improve hydrological and meteorological forecasting

- Enhance disaster preparedness
- Improve policies and legal frameworks for regulation of water resources
- Establish state regulatory agencies and river basin authorities and build their capacity
- Clean-up rivers
- Build capacity of water user associations to manage irrigation systems
- Implement state Sector-Wide Approaches (SWAps) governing investments in and management of rural water and sanitation schemes
- Demonstrate sustainable approaches to development and management of urban water supply
- Recent attention to climate change is also encouraging. A number of studies are being carried out to analyze the impacts of climate change at the river basin, sub-basin, city and community level as the basis for developing strategies for adaptation. In addition, the private sector is conducting water audits and developing strategies to improve efficiency, reduce their "water footprint", and enhance the sustainability of water resources.

IRRIGATION IN INDIA

Irrigation in India includes a network of major and minor canals from Indian rivers, groundwater well based systems, tanks, and other rainwater harvesting projects for agricultural activities. Of these groundwater system is the largest. In 2013-14, only about 47.7% of total agricultural land in India was reliably irrigated. The largest canal in India is Indira Gandhi Canal, which is about 650 km long. About 2/3rd cultivated land in India is dependent on monsoons. Irrigation in India helps improve food security, reduce dependence on monsoons, improve agricultural productivity and create rural job opportunities. Dams used for irrigation projects help produce electricity and transport facilities, as well as provide drinking water supplies to a growing population, control floods and prevent droughts.

Types of Irrigation

Depending upon the availability of surface or ground water, topography, soil and rivers, various types of irrigation practised in India are as follows:

Tanks

In the Deccan, water-reservoirs are made by constructing dams. This system is greatly adopted in the States of Tamil Nadu, Andhra Pradesh, and Karnataka, etc. In Northern India also, tanks are constructed for storing water. From all these tanks, water is carried to the fields through canals.

In many places, rain-water harvesting systems are installed and water is stored in large artificial reservoirs to be used for agricultural purposes.

It is prevalent in the uneven and relatively rocky plateau of Peninsular India.

Tanks are commonly used in Deccan Plateau, Andhra Pradesh, Karnataka, Tamil Nadu and Maharashtra. About 8% of total irrigated area is irrigated by tanks.

Most of the tanks are small in size and built by individuals or group of farmers by raising bunds across seasonal streams. But there are some drawbacks: Tanks cover a large areas of cultivable land. Evaporation of water is rapid due to large expanse of shallow water of tanks, do not ensure perennial supply of water.

Well

It is more widespread in plains, coasts and some regions of peninsular India. It is less costly and more flexible as water can be drawn whenever needed and 'evaporation loss' is minimised and no fear of "over irrigation". Uttar Pradesh leads in well irrigation and is followed by Punjab, Haryana, Bihar, Gujarat and Andhra Pradesh.

Wells are mainly found in U. P., Bihar, Tamil Nadu, etc. there are various types of wells - shallow wells, deep wells, tube wells, artesian wells, etc. From the shallow wells water is not always available as the level of water goes down during the dry months. Deep wells are more suitable for the purpose of irrigation as water from them is available throughout the year.

Tube-wells are also used for irrigation purposes. At places where ground water is available, a tube-well can be installed near the agricultural area. A deep tube well worked by electricity, can irrigate a much larger area (about 400 hectares) than a surface well (1/2 hectares). Tube wells are mostly used in U.P., Haryana, Punjab, Bihar and Gujarat. In Rajasthan and Maharashtra, artesian wells are now supplying water to agricultural lands. In artesian wells, water level remains at a high-level because of the natural flow of water due to high pressure.

Wells are of two types

Open wells and tube-wells. Open wells are shallow and irrigate a small area because water available is limited. Tube wells are deep and have the capacity to draw a large volume of water. It has increased in recent years.

Canals

Canals are the main source of irrigation in India. Canals are big water channels taken out from rivers to carry water to places far away from the river.

Canal irrigation is playing a vital role in Indian agriculture. It canal near about 42% of total irrigated land. In many places during the rainy season, there is flood in the rivers. The flood water is carried to the field through canals. These canals are found in W.B., Bihar, Orissa, etc. They supply water only when there is flood in the rivers, and therefore, are of no use during the dry season when water is required most.

It is of two types: Canals taken out from rivers without any regulating system like weirs etc. at their head are called inundation canals and canals taking off from perennial rivers with a weir system to regulate flow of water are called perennial canals.

Punjab and Haryana have become the first granaries of country due to these canals which include Western Yamuna Canal, Sirhind Canal, Upper Bari Doab canal & Bliakra Canal. The important canals of Uttar Pradesh are upper and lower Ganga Canal, Agra and Sharda Canal and Rajasthan has become third granary due to Rajasthan canal project.

PERENNIAL CANALS IRRIGATION SYSTEM

The perennial canals get the supply of water either from the river directly or through the reservoirs of the river projects. In order to supply water throughout the year, reservoirs are constructed for storing water. From these reservoirs, water can be supplied to the fields whenever there is demand for it. So this system of irrigation ensures supply of water in all season. This type of perennial canal is found mostly in Punjab, U.P., and Tamil Nadu.

In Punjab, the upper Bari Doab canal connecting the Ravi and the Beas and Sirhind (from the Sutlej) canal is famous. In U.P., the Upper Ganga and the Lower Ganga canals, Agra and Sarda canals, etc. are important. In Tamil Nadu, most important are the Buckingham canal and the Periyar canal.

MULTI PURPOSE RIVER VALLEY PROJECTS

In recent years, multi-purpose river valley projects are helping agriculture. The most important are:

- the Damodar Valley Project in West Bengal,
- the Mor (Mayurakshi) Project in West Bengal,
- the Mahanadi (Hirakud) Project in Orissa,
- the Kosi Project in Bihar, and
- the Bhakra Nangal Project in Punjab.

Methods of irrigation

Efficient methods of irrigation enable the application of the right amount of water to the crop at the right time and its uniform distribution in the field. Irrigation water may be applied to the crops by spreading it in a sheet or in small streams on the soil surface, by spraying it over the crop through pipes and nozzles or by applying it in drops to wet the crop root zone.

Border irrigation

Border irrigation is a controlled surface flooding method of water application. The field is divided into a number of long parallel strips called borders, that are separated by low ridges. The width of a border usually varies from 3 to 15 m. The length of a border varies from 60 to 120 m in sandy and sandy loam soil, 100 to 180 m in medium loam soils and 150 to 300 m in clay loam or clay soils. The border method of irrigation is most suitable for close growing grain crops, such as wheat, barley and fodder crops. Field trials in India have shown that border irrigation is suitable in most areas of the country. Water is retained in the basin until it soaks into the soil.

Basin irrigation

The basin method of irrigation is essentially the check basin method applied to orchards. In flat lands, the basins are square, while on sloping lands the basins are formed by constructing ridges along adjacent contours with short cross ridges at either end. Usually there is one tree to a basin.

Sprinkler irrigation

In sprinkler irrigation, water is sprayed into the air through a sprinkler nozzle and allowed to fall on the land surface in a uniform pattern at a rate less than the infiltration rate of the soil. Sprinklers were introduced in India during the early 1950s. Initially, the sprinklers were used on high value plantation crops such as tea, coffee, chicory, cardamom and in orchards. Their use is gaining popularity on food crops, orchards, cotton and vegetables in areas where sprinklers are economically justifiable and technically feasible. Sprinkler irrigation can be used for almost all crops (except rice and jute) and on nearly all soils.

Drip irrigation

Drip irrigation, also called trickle irrigation, involves slow application of water to the plant root zone. The losses by deep percolation and evaporation are minimized. A precise amount of water is applied to replenish the depleted soil moisture at frequent intervals, for optimum plant growth. The system enables the application of water and fertilizer at an optimum rate to the plant root system. The amount of water supplied to the soil is almost equal to the daily consumptive use, thus maintaining a low moisture tension in soil.

Future gains in agricultural productivity of the country shall be critically determined by proper development and utilization of its surface and groundwater resources. Technological developments in pump manufacturing and deep well drilling have made groundwater extraction a common phenomenon, and the last three decades have witnessed a substantial increase in the groundwater development and utilization. Presently, groundwater is used to irrigate almost 50 percent of the total irrigated land in the country through about 5.5 million tubewells and more than 10 million dug wells. Unfortunately, these advantages have led to overexploitation and unsustainable utilization of this precious natural resource.

Need for Irrigation

(i) India is a big country and stands next to China when we talk about population and so irrigation facilities are needed to grow more food to feed our teaming millions.

(ii) The distribution in rainfall is uneven and uncertain which either causes famines or drought. By means of irrigation we can check both the problems.

(iii) Different water requirements of different crops can only be met through irrigation facilities.

(iv) India, being a tropical country the temperature is high and evaporation more rapid, so, artificial irrigation is necessary for ample supply of water and also to prevent water scarcity in the long dry winter season.

IMPACT OF IRRIGATION ON HUMAN ENVIRONMENT

An irrigation project, like any other water resource project, is constructed for the well-being of people, and has definite impact on the environment. The developmental activities cannot be stopped on environmental considerations alone. One should, however, appreciate that both developmental activities and an intact environment are equally important for the sustained well-being of people.

Following are die advantages of providing irrigation in an area:

(i) It provides protection against famine.

(ii) It increases agricultural production and its quality and, thus, adds to the national economy by substituting imports, and generating exports of food and non-food agricultural products.

(iii) It reduces rural poverty by creating additional employment in the sectors of main agricultural activities and related activities of input supplies and processing and marketing of agricultural products.

(iv) Inspection roads constructed for the irrigation works can be used by the people of the area and, thus, irrigation helps in improving the communication system of the region.

(v) Irrigation works can also be used for hydropower generation, inland navigation, industrial and domestic water supply, flood control, etc.

As a result of these benefits of irrigation, there are permanent gains in the life-style of rural population due to increased prosperity. However, there are some disadvantages associated with irrigation facilities which cause considerable increase in water-borne and water-related diseases.

Excessive surface water irrigation can lead to waterlogging of the cultivable land making it unsuitable for cultivation. There can be some adverse effects on the environment too which may adversely affect the human beings in the long run. With proper planning, these disadvantages can, however, be minimised, if not eliminated completely.

MANAGEMENT OF POOR QUALITY WATER

Selection of crops, their varieties and salt tolerance at various stages of their growth are also important. Detailed guidelines for using saline irrigation water in India are available.

In the extreme cases of waterlogging and salinity, the seeds may not germinate and the plants may wilt permanently. Waterlogging and salinity of agricultural lands are due to a natural process or man-made activities. In irrigated areas, chemical degradation of land is a subsequent development on account of long duration waterlogging. Besides the erratic rainfall distribution, inundation of coastal areas by back-water flow from the sea and absence of proper soil and water-conservation measures in the catchments are also important factors leading to drainage congestion and related problems in the agricultural lands.

There is a lack of up-to-date and methodically monitored information on the type, severity and extent of drainage problems in the country. Estimates of waterlogged areas in India range from 2.5 to 16 mha, and those for salinized areas from 3.3 to 10.9 mha.

Beyond the above gross estimates for the whole country, very few published records based on rigorous monitoring exist on the progressive development of waterlogging and salinity in India.

Management of rain water resources

The technologies of rain water management are highly location-specific and are determined by physiographic, environmental, technical and socio-economic reasons. The main features are to conserve rain water where it falls, i.e. in the soil profile up to its maximum capacity and in storage structures for crop irrigation at a later stage. The technologies practised are (i) in situ water conservation, (ii) supplemental irrigation by run-off recycling, (iii) watershed management including agriculture, (iv) pond culture adjunct to canal system, (v) construction of high tidal-dykes, (vi) rice-fish integrated farming, (vii) using pressurized irrigation system, (viii) groundwater management, (ix) flood water management, (x) management of drainage and water-logging. Needless to say, ultimately the approaches have to be compatible with socio-economic adjustments of the target population.

Micro-irrigation in India

Micro-irrigation is the slow application of water as discrete or continuous drips, tiny streams or miniature spray on, above, or below the soil by surface drip, subsurface drip, bubbler and micro-sprinkler systems. It is applied through emitters connected to a water delivery line through low-pressure delivery.

As the agriculture sector consumes 80% of the freshwater in India, micro-irrigation is often promoted by central and state governments as a way to tackle the growing water crisis. This is because drip and sprinkler irrigation delivers water to farms in far lesser quantities than conventional gravity flow irrigation. Due to recurring droughts in years 2012, 2015 and 2016, micro-irrigation has become a policy priority in India. The new catch-phrase in one of the central government's schemes, Pradhan Mantri Krishi Sinchai Yojana (PMKSY or Prime Minister's Agriculture Irrigation Programme), is "Per Drop More Crop". Apparently, the shift towards micro-irrigation is thought to "save" water and boost crop yields.

But, an investigation into the realities across India's farms, basins, and sectors linked with water resources reveals that micro-irrigation program in India is fraught with a risk of failure. This is because the program is ill-conceived, not holistic, and disintegrated from the basin/watershed perspective.

So questions arise: What are the risks associated with micro-irrigation in India? More importantly, under the current set of assumptions can micro-irrigation really "save" water and boost crop yields? - the premise over which the entire scheme is built and being promoted in India. This article therefore attempts to outline, assess, and reveal the risks and bottlenecks associated with micro-irrigation in India, so that course correction can be undertaken in future.

Limitations of Irrigation on Account of Certain Problems Related to Irrigation:

Despite large-scale investment and expansion of irrigation facilities, it is a matter of serious concern that about 60 per cent of the total cropped area is still dependent on rain. There are a number of problems related to irrigation and they have to be solved.

(1) Delays in completion of projects:

The biggest problem in our major and medium irrigation sectors right from the First Five Year Plan has been the tendency to start more and more new projects resulting in wanton proliferation of projects. There is also delay in utilisation of potentials already present. In most of the projects, there have been delays in construction of field channels and water courses, land leveling and land shaping.

(2) Inter-state water disputes:

Irrigation is a state subject in India. Development of water resource is, therefore, being planned by states individually taking into account their own needs and requirement. However, all major rivers are inter-state in character. As a result, differences with regard to storage, priorities and use of water arise between different states. Narrow regional outlook brings inter-state rivalries over distribution of water supply.

(3) Regional disparities in irrigation development:

The Ninth Five Year Plan Document estimated that the water resource development in North Eastern region through major, medium and minor schemes is only at the level of 28.6 per cent whereas in the Northern region it has reached about 95.3 per cent. This indicates a wide regional variation in the development of irrigation facilities.

(4) Waterlogging and salinity:

Introduction of irrigation has led to the problem of waterlogging and salinity in some of the states. The working group constituted by the Ministry of Water Resources in 1991 estimated that about 2.46 million hectares in irrigated commands suffered from water logging. The working group also estimated that 3.30 million hectares had been affected by salinity/alkalinity in the irrigated commands.

(5) Increasing cost of irrigation:

The cost of providing irrigation has been increasing over the years from the First Five Year Plan to Tenth First Five Year Plan.

(6) Losses in operating irrigation projects:

While just prior to Independence (1945-46) public irrigation schemes showed a surplus after meeting working expenses and other charges. The position deteriorated considerably in the post-Independence period.

(7) Decline in water table:

There has been a steady decline in water table in the recent period in several parts of the country, especially in the western dry region, on account over exploitation of ground water and insufficient recharge from rain water.

DEVELOPMENT AND SCOPE OF IRRIGATION IN INDIA

Irrigation has been practiced throughout the world since the beginning of civilisation. In India, the Grand Anicut across the river Cauvery was constructed in the second century. There were a large number of water tanks in peninsular India and several inundation canals (such canals draw their supplies from a river only during the high stages of the river) in northern India.

The Upper Ganga canal, the Upper Bari Doab canal and the Krishna and the Godavari delta systems were built between 1836 and 1866. The famines of 1876-78, 1897-98 and 1899-1900 led to the setting up of the first Irrigation Commission in 1901.

In 1901, the total gross irrigated area was only 13.3 Mha (million hectares) which increased to 22.6 Mha in 1950 as a result of a spurt in protective irrigation schemes. The Bengal famine of 1943 made the authorities appreciate the urgency of increasing agricultural production to meet the needs of the growing population.

After independence in 1947, an era of planned development in irrigation started with the first five-year Plan in 1951. Giant projects like the Bhakra-Nangal, Hirakud, Damodar valley, Nagarjunasagar, Rajasthan canal, etc., were taken up.

As a result, the irrigated area increased from 22.6 Mha in 1951 to 87.8 Mha in 1995. The total ultimate irrigation potential is estimated at 115.54 Mha of which 58.47 Mha would be from major and medium irrigation schemes, and the remaining from minor irrigation schemes.

Command Area Development

The irrigation potential created by the construction of a large number of irrigation projects in India has almost quadrupled since independence. It has, however, always remained under-utilised, and the optimum benefits by way of increased production have not been fully realised. Against the world average of about 5 tonnes of agricultural production per hectare of irrigated land, India produces only about 2.2 tonnes of crop per hectare of irrigated land.

The following have been identified as the main causes for these deficiencies:

(i) Need for modernisation of the old irrigation systems.

(ii) Lack of adequate drainage which results in waterlogging of the cultivable land.

(iii) Inadequate on-farm management of water.

(iv) Lack of suitable infrastructure and extension services.

(v) Poor coordination amongst the concerned organisations in the command areas.

Gross command area (or command area) of an irrigation system is the total area which can be economically irrigated from the system without considering the limitations on the quantity of available water. For optimising the benefits of available irrigation potential, an integrated command area development is always recommended.

The objectives of the command area development programme are as follows:

1. Increasing the area of irrigated land by proper land development and water management.
2. Optimising yields by adopting the best cropping pattern consistent with the availability of water, soil and other local conditions.
3. Bringing water to the farmer's fields rather than only to the outlets and, thus, assuring equitable distribution of water and adequate supply to tail-enders.
4. Avoiding wastage and misuse of water.
5. Optimising the use of scarce land and water resources, including ground water where available, in conjunction with necessary inputs and infrastructure.

The command area development programme includes the following components:

1. Modernisation and efficient operation of the irrigation system.
2. Development of main drainage system.
3. Construction of field channels and field drain.
4. Land shaping/levelling and consolidation of land holdings.
5. Lining of field channels/watercourses.
6. Exploitation of ground water and installation of tube wells.
7. Adoption and enforcement of a suitable cropping pattern.
8. Enforcement of an appropriate rostering system on irrigation.
9. Making arrangement for timely supply of key inputs like credit, seeds, fertilisers, pesticides and implements.
10. Strengthening of existing extension, training and demonstration or-ganisations.

3

Irrigation Practices in India

Being a agriculture dependent country, irrigation is the backbone of India. India is a vast country with a kaleidoscopic diversity if topography, climate and vegetation. The rainfall is generally capricious in its incidence and variable in amount. The distribution of water in India is therefore, very uneven. The rainfall in this country is concentrated usually during four months in a year when there is excess water which flows down unutilized, while in other seasons there is acute shortage of water.

The total of cultivable area in this country is about 185 million hectares. At present about 172 million hectares are under cultivation. Seventy percent of India's vast population depends upon agriculture directly for their living, and therefore agriculture has always been and promises to remain the main industry of India in foreseeable future also. India has large water resources, great rivers systems and vast thirsty tracts of land and is thus designed, so to say by nature for the development of irrigation. This is why India has some of the earliest irrigation works.

However, even when full potential of available resources are developed, irrigation facilities can be extended to 115 million ha of which 80 million ha from surface water and 35 million ha from ground water. The gross cropped area is expected to increase to about 200 million ha during the next two decades due to introduction of multiple cropping and land reclamation.

HISTORY OF IRRIGATION DEVELOPMENT IN INDIA

Ministry of Water Resources (Govt. Of India), on its web site briefly explains the history of irrigation development in India which can be traced back to prehistoric times. Vedas, Ancient Indian writers and ancient Indian scriptures have made references to wells, canals, tanks and dams. These irrigation technologies were in the form of small and minor works, which could be operated by small households to irrigate small patches of land. In the south, perennial irrigation may have begun with construction of the Grand Anicut by the Cholas as early as second century to provide irrigation from the Cauvery River. The entire landscape in the central and southern India is studded with numerous irrigation tanks which have been traced back to many centuries before the beginning of the Christian era. In northern India also there are a number of small canals in the upper valleys of rivers which are very old.

Ghiyasuddin Tughluq (1220-1250) is credited to be the first ruler who encouraged digging canals. Fruz Tughlug (1351-86) is considered to be the greatest canal builder. Irrigation is said to be one of the major reasons for the growth and expansion of the Vijayanagar Empire in southern India in the fifteenth century. Babur, in his memoirs called 'Baburnamah' gave a vivid description of prevalent modes of irrigation practices in India at that time. The Gabar Bunds captured and stored annual runoff from surrounding mountains to be made available to tracts under cultivation.

Close to nineteenth century according to sources of irrigation; canals irrigated 45 %, wells 35 %, tanks 15 % and other sources 5 %. Famines of 1897-98 and 1899-1900 necessitated British to appoint first irrigation commission in 1901, especially to report on irrigation as a means of protection against famine in India. As a result of recommendations of first irrigation commission total irrigated area by public and private works increased to 16 Mha in 1921. From the beginning of 19th century to 1921 there was no significant increase in tube well irrigated area. During 1910 to 1950 growth rate of irrigation was estimated at 2.0 % per annum for government canal irrigation, 0.54 % per annum for well irrigation and 0.98 % per annum in respect of irrigation from all sources.

At time of independence net irrigated area of India under British rule which include Bangladesh and Pakistan was 28.2 Mha. After partition net irrigated area in India and Pakistan being 19.4 Mha and 8.8 Mha respectively.

Different types of irrigation and irrigation systems storage

Agriculture is the nerve of any country as it is needed for survival of living beings. For growing crops, irrigation is major process. Irrigation is described as the artificial application of water to the land or soil. It is the substitute or supplement of rainwater with another source of water. It is used in dry areas and during periods of insufficient rainfall. It is considered as basic infrastructure and vital input required for agricultural production (Mamata Swain, 1999). Major aim of irrigation systems is to help out in the growing of agricultural crops and vegetation by maintaining with the minimum amount of water required, maintenance of landscapes, and re-vegetation of disturbed soils. Irrigation systems are also used for dust repression, removal of sewage, and in mining. Irrigation is often studied together with drainage, which is the natural or artificial removal of surface and sub-surface water from a given region. Agricultural scientists stated that irrigation also has a few other uses in crop production, which include protecting plants against frost, suppressing wild plant growing in grain fields and helping to avert soil consolidation. On the contrary, agriculture that relies only on direct rainfall is referred to as rain-fed or dry-land farming.

Irrigation has been a vital characteristic of agriculture for over centuries and the result of work of many cultures, and was the basis of the wealth and society ranging from Asia to the American Southwest.

In brief, irrigation also has many applications in crop production, which include:

1. Protecting plants against frost
2. Suppressing weed growth in grain fields
3. Preventing soil consolidation.
4. for dust suppression,
5. Disposal of sewage, and in mining.

Irrigation water may be applied to the crops in three basic methods that include surface irrigation, sprinkler irrigation and sub-surface irrigation method. Every irrigation method has advantages and disadvantages. Therefore before choosing a specific technique, the irrigation engineer must evaluate all the factors and choose method which is most suited for local condition (Dr. B. C. Punmia, 2009).

TYPE OF IRRIGATION TECHNIQUE IN INDIA

In India, the irrigated area consists of about 36 per cent of the net sown area. There are various techniques of irrigation practices in different parts of India. These methods of irrigation differ in how the water obtained from the source is distributed within the field. In general, the goal of irrigation is to supply the entire field homogeneously with water, so that each plant has the amount of water it needs, neither too much nor too little. Irrigation in India is done through wells, tanks, canals, perennial canal, and multi-purpose river valley projects.

Surface Irrigation

Surface irrigation has evolved into an extensive array of configurations which can be broadly classified as: (1) basin irrigation; (2) border irrigation; (3) furrow irrigation; and (4) uncontrolled flooding. As noted previously, there are two features that distinguish a surface irrigation system: (a) the flow has a free surface responding to the gravitational gradient; and (b) the on-field means of conveyance and distribution is the field surface itself.

A surface irrigation event is composed of four phases as illustrated graphically in Figure 1. When water is applied to the field, it 'advances' across the surface until the water extends over the entire area. It may or may not directly wet the entire surface, but all of the flow paths have been completed. Then the irrigation water either runs off the field or begins to pond on its surface. The interval between the end of the advance and when the inflow is cut off is called the wetting or ponding phase. The volume of water on the surface begins to decline after the water is no longer being applied. It either drains from the surface (runoff) or infiltrates into the soil. For the purposes of describing the hydraulics of the surface flows, the drainage period is segregated into the depletion phase (vertical recession) and the recession phase (horizontal recession). Depletion is the interval between cut off and the appearance of the first bare soil under the water. Recession begins at that point and continues until the surface is drained.

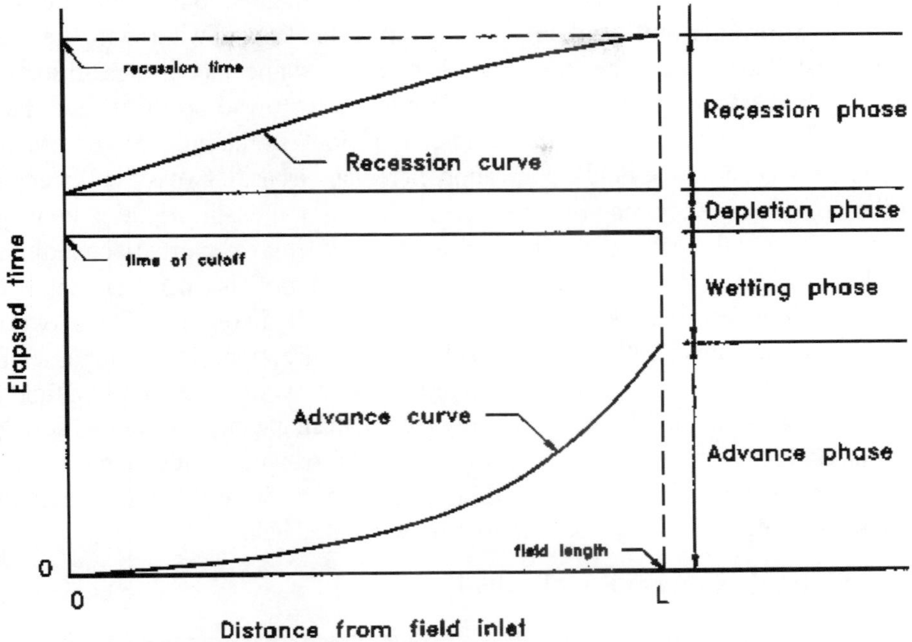

Figure 1: Time-space trajectory of water during a surface irrigation showing its advance, wetting, depletion and recession phases.

The time and space references shown in Figure 1 are relatively standard. Time is cumulative since the beginning of the irrigation, distance is referenced to the point water enters the field. The advance and recession curves are therefore trajectories of the leading and receding edges of the surface flows and the period defined between the two curves at any distance is the time water is on the surface and therefore also the time water is infiltrating into the soil.

It is useful to note here that in observing surface irrigation one may not always observe a ponding, depletion or recession phase. In basins, for example, the post-cut off period may only involve a depletion phase as the water infiltrates vertically over the entire field. Likewise, in the irrigation of paddy rice, an irrigation very often adds to the ponded water in the basin so there is neither advance nor recession - only wetting or ponding phase and part of the depletion phase. In furrow systems, the volume of water in the furrow is very often a small part of the total supply for the field and it drains rapidly. For practical purposes, there may not be a depletion phase and recession can be ignored. Thus, surface irrigation may appear in several configurations and operate under several regimes.

This divided into the four component systems: (1) water supply; (2) water conveyance or delivery; (3) water use; and (4) drainage.

Basin irrigation

Basin irrigation is the most common form of surface irrigation, particularly in regions with layouts of small fields. If a field is level in all directions, is encompassed by a dyke to prevent runoff, and provides an undirected flow of water onto the field, it is herein called a basin. A basin is typically square in shape but exists in all sorts of irregular and rectangular configurations. It may be furrowed or corrugated, have raised beds for the benefit of certain crops, but as long as the inflow is undirected and uncontrolled into these field modifications, it remains a basin. Two typical examples are shown in Figure 2, which illustrate the most common basin irrigation concept: water is added to the basin through a gap in the perimeter dyke or adjacent ditch.

Basin irrigation is common practice of surface irrigation. This method is employed for watering orchards (Basak, 1999). It is useful especially in regions with layouts of small fields (Shah et al. 2002). If a field is level in all directions, is encompassed by a dyke to prevent runoff, and provides an undirected flow of water onto the field, it is herein called a basin. A basin is typically square in shape but exists in all sorts of irregular and rectangular configurations. It may be furrowed or ridged, have raised beds for the benefit of certain crops, but as long as the inflow is undirected and uncontrolled into these field modifications, it remains a basin.

Basin Method (Source: Basak, 1999)

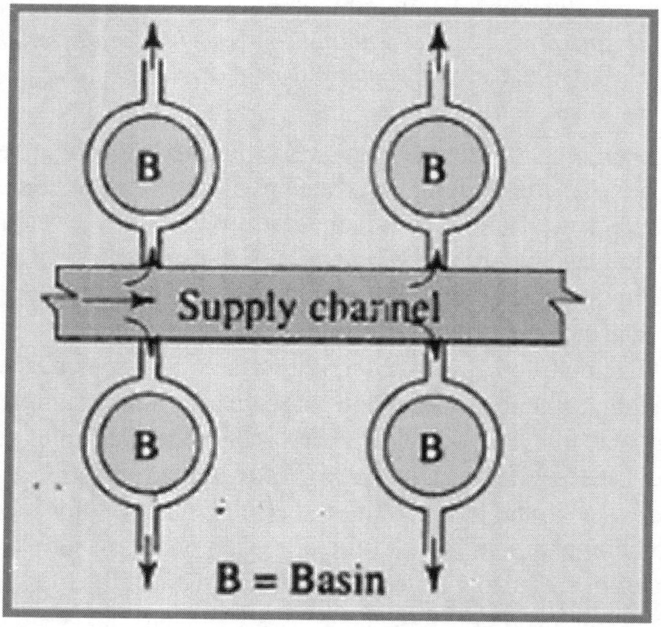

Furrow Irrigation

Furrow irrigation avoids flooding the entire field surface by channelling the flow along the primary direction of the field using 'furrows,' 'creases,' or 'corrugations'.

Water infiltrates through the wetted perimeter and spreads vertically and horizontally to refill the soil reservoir. Furrows are often employed in basins and borders to reduce the effects of topographical variation and crusting. The distinctive feature of furrow irrigation is that the flow into each furrow is independently set and controlled as opposed to furrowed borders and basins where the flow is set and controlled on a border by border or basin by basin basis.

Furrows provide better on-farm water management flexibility under many surface irrigation conditions. The discharge per unit width of the field is substantially reduced and topographical variations can be more severe. A smaller wetted area reduces evaporation losses. Furrows provide the irrigator more opportunity to manage irrigations toward higher efficiencies as field conditions change for each irrigation throughout a season. This is not to say, however, that furrow irrigation enjoys higher application efficiencies than borders and basins.

There are several disadvantages with furrow irrigation. These may include: (1) an accumulation of salinity between furrows; (2) an increased level of tailwater losses; (3) the difficulty of moving farm equipment across the furrows; (4) the added expense and time to make extra tillage practice (furrow construction); (5) an increase in the erosive potential of the flow; (6) a higher commitment of labour to operate efficiently; and (7) generally furrow systems are more difficult to automate, particularly with regard to regulating an equal discharge in each furrow. Figure 3 shows two typical furrow irrigated conditions.

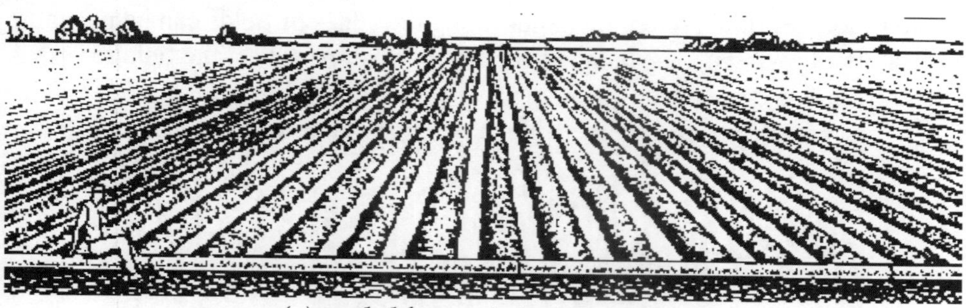

(a) graded furrow irrigation system

(b) contour furrows

Figure 3. Furrow irrigation configurations (after USDA-SCS, 1967)

In different situations, different furrow methods are used (Surajbhan 1978). They are mainly of five types:

1. Sloppy Furrow
2. Levelled Furrow
3. Contour Furrow
4. Serial Furrow
5. Corrugated Furrow

There are numerous advantages of Furrow technique of irrigation:

1. Large areas can be irrigated at a time.
2. It saves labour since once the furrow is filled, it is not necessary to give water a second time.
3. It is a reasonably cheaper method.
4. Plants get proper quantity of water by this system.

Furrow irrigation is also beneficial for growing of tree crops. In the early stages of tree planting, one furrow alongside the tree row may be sufficient but as the trees develop then two or more furrows can be constructed to provide sufficient water. Sometimes a special zig-zag system is used to improve the spread of water (Basak, 1999).

Major drawback of furrow system of irrigation is ensuring uniform dispersal of water over a given field. To tackle this problem, some farmers engage in field levelling to remove any small hills that would have been bypassed by the gravity flow of the water. Other problem with furrow irrigation is the increased potential for water loss due to runoff. Building retention ponds along the edges of fields can help capture this runoff, allowing it to be pumped back to the upslope side of the field for use in further irrigation cycles.

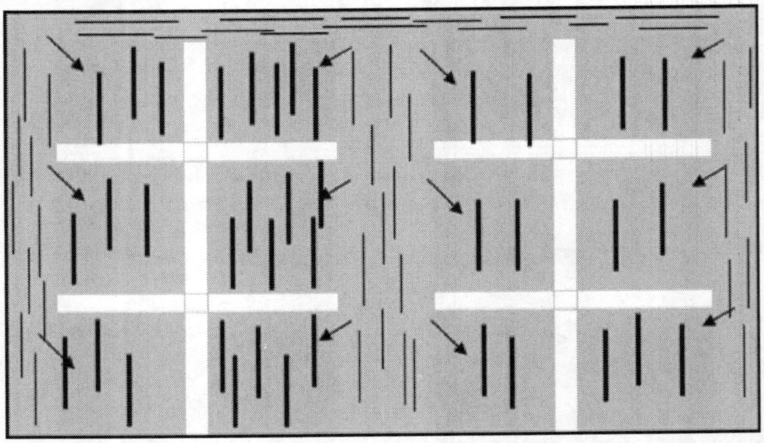

Uncontrolled flooding: There are many cases where croplands are irrigated without regard to efficiency or consistency. These are usually situations where the value of the crop is very small or the field is used for grazing or recreation purposes. Small land holdings are generally not subject to the range of surface irrigation practices of the large industrial farming systems. The assessment methods can be applied if desired, but the design techniques are not generally applicable nor need they be since the irrigation practices tend to be minimally managed.

Free Flooding

There are many cases where croplands are irrigated without regard to efficiency or uniformity. These are generally situations where the value of the crop is very small or the field is used for grazing or recreation purposes. Small land holdings are generally not subject to the array of surface irrigation practices of the large commercial farming systems. Also in this category are the surface irrigation systems like check-basins which irrigate individual trees in an orchard, for example. While these systems represent significant percentages in some areas, they will not be discussed in detail in this paper. The evaluation methods can be applied if desired, but the design techniques are not generally applicable nor need they be since the irrigation practices tend to be minimally managed.

This flooding system of irrigation is used from ancient times. Flooding method consists in applying the water by flooding the land of rather smooth and flat topography. In current irrigation practice, several flooding methods have been developed. In free flooding method, water is applied to the land from field ditches without any check or guidance to the flow. The land is divided into plots or kiaries of suitable size depending on porosity of soil. Water is spread over the field from watercourse. The irrigation operation begins at the higher area and proceeds towards the lower levels. The flow is stopped when the lower end of the field has received the desired depth of water. The field watercourse is properly spaced, the spacing depends on the topography, oil texture, depth of soil and size of stream.

Free Flooding for erodible soil

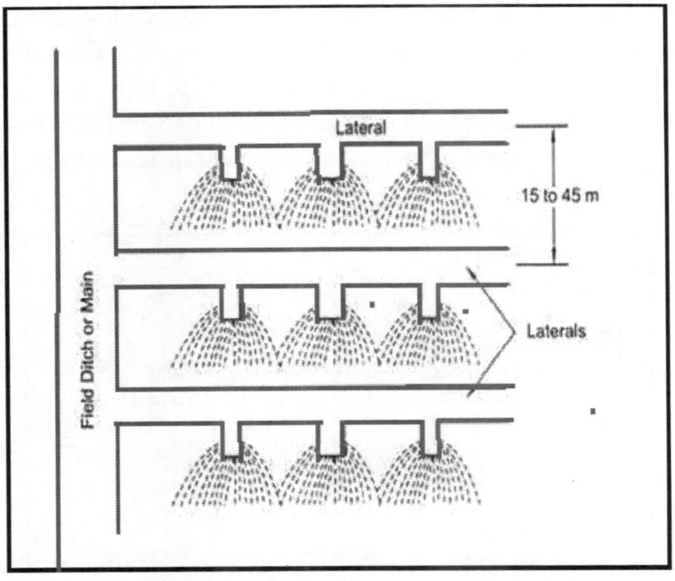

This technique is beneficial for newly established farms where making furrows is very expensive. This method is economical and can be effectively used where water supply is in plenty. This method is suitable for the fields with irregular surface in which other techniques are difficult to apply. Major drawback of this method is that there is no perfect control over the flow of water to attain high efficiency. Sometimes the flow of water over the soil is too rapid to fulfil soil moisture deficiency. On the other hand, sometimes water is retained on the field for a very long time and consequently the water is lost in infiltration or deep percolation.

Border Strip Method

Border irrigation can be viewed as an extension of basin irrigation to sloping, long rectangular or contoured field shapes, with free draining conditions at the lower end. Figure 4 illustrates a typical border configuration in which a field is divided into sloping borders. Water is applied to individual borders from small hand-dug checks from the field head ditch. When the water is shut off, it recedes from the upper end to the lower end. Sloping borders are suitable for nearly any crop except those that require prolonged ponding. Soils can be efficiently irrigated which have moderately low to moderately high intake rates but, as with basins, should not form dense crusts unless provisions are made to furrow or construct raised borders for the crops. The stream size per unit width must be large, particularly following a major tillage operation, although not so large for basins owing to the effects of slope. The precision of the field topography is also critical, but the extended lengths permit better levelling through the use of farm machinery.

In this technique of irrigation, a field is divided into number of strips. The width of strip varies from 10 to 15 metres and length varies from 90 m to 400 m. Strips are separated by low embankments or levees. The water is diverted from the field channel into the strips. The water flows gradually towards lower end, wetting the soil as it advances. The surface between two embankments should essentially be level. It assists in covering the entire width of the strip. There is a general surface slope from opening to the lower end. The surface slope from 2 to 4 m/1000 m is best suited. When the slope is steeper, special arrangement is made to prevent erosion of soil.

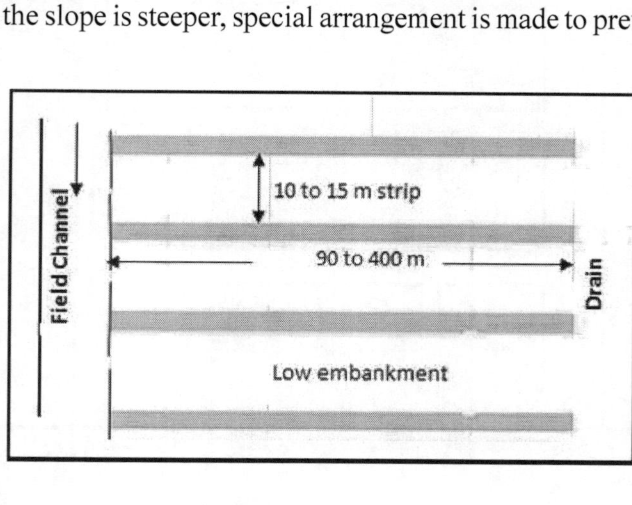

Well and Tube Well Irrigation

A well is a hole dug in the ground to get the subsoil water. Normal well is about 3-5 metres deep but deeper wells up-to 15 metres are also dug. This system of irrigation has been used in India from ancient time. Various methods are used to lift the ground water from the well for irrigation, drinking, bathing and for other purposes. Well irrigation is more popular in those regions where ground water is in ample and where there are few canals. These areas include a large part of the Great Northern Plain, the deltaic regions of the Mahanadi, the Godavari, the Krishna and the Cauvery, parts of the Narmada and the Tapi valleys and the weathered layers of the Deccan Trap and crystalline rocks and the sedimentary zones of the Peninsula. However, the greater part of the Penisnular India is not appropriate for well irrigation due to stony structure, rough surface and lack of underground water. Large dry tracts of Rajasthan, the adjacent parts of Punjab, Haryana, and Gujarat and some parts of Uttar Pradesh have salty ground water which is not suitable for irrigation and human use and hence inappropriate for well irrigation.

Tube well: A tube well is a deeper well (generally over 15 metres deep) from which water is lifted with the help of a pumping set operated by an electric motor, a diesel engine or solar power. A tube well cannot be constructed in all places and requires some geographical conditions favouring its installation. The main factors for Tube well construction are:

1. There should be enough quantity of ground water because a tube well can generally irrigate 2 hectares per day against 0.2 hectares per day irrigated by an ordinary well.
2. The water level should be nearly 15 metres. If the water table is more than 50 metres deep the cost of pumping out water from the tube well becomes uneconomic.
3. There should be normal supply of cheap electricity or diesel so that water from the tube well can be taken out at the hour of need.
4. The soil in the immediate neighbourhood of the tube-well should be productive so that there is demand for irrigation and the cost involved in the construction and operation of the tube well can be recovered by the increased farm production.

Advantages of Well and Tube Well Irrigation:

Well is simplest and cost effective source of irrigation and the poor Indian farmer can easily afford it. Well is an independent source of irrigation and can be used as and when the necessity arises. Several chemicals such as nitrate, chloride, sulphate, etc. are generally found mixed in well water. They make soil fertility when they reach the agricultural field along with well water. It is more reliable during periods of famine when surface water dries up. Disadvantages of well and Tube Well Irrigation: With these methods, only limited area can be irrigated. Normally, a well can irrigate 1 to 8 hectares of land. The well may dry up and may be rendered ineffective for irrigation if excessive water is taken out. In the drought situation, the ground water level falls and enough water is not available in the well when it is needed the most. Tube wells can draw a lot of groundwater from its neighbouring areas and make the ground dry and not suitable for agriculture. Well and tube well irrigation is not possible in areas of salty groundwater.

Canal Irrigation

Canals are most important source of irrigation from the period of 1960s, but in the 1970s, they yielded first place to wells and tube wells and now constitute the second most important source of irrigation in India. Canals are most effective techniques of irrigation in areas of low level relief, deep fertile soils, perennial source of water and extensive command area. Therefore, the main concentration of canal irrigation is in the northern plain of India, especially the areas comprising Uttar Pradesh Haryana and Punjab.

The digging of canals in stony and uneven areas is difficult and unprofitable. Thus the canals are practically absent from the Peninsular plateau area. However, the coastal and the delta regions in South India do have some canals for irrigation.

CANALS IN INDIA ARE OF TWO TYPES

Inundation canals, which are taken out from the rivers without any regulating system like weirs at their head. Such canals supply irrigation mainly in the rainy season when the river is in flood and there is excess water. When the rainy season is over, the flood in the river subsides, the level of water falls below the level of the canal head and the canal dries up. Some canals taken off from the Satluj in Punjab were of this type. Since irrigation from this type of canals is unsure, they have been converted in perennial canals.

Perennial Canals are those which are taken off from perennial rivers by constructing a barrage across the river. Most of the canals in India today are perennial.

Advantages of Canal Irrigation: Most of the canals provide perennial irrigation and supply water as and when required. This saves the crops from drought conditions and helps in increasing the farm production. Canals carry a lot of residue brought down by the rivers. This sediment is deposited in the agricultural fields which make soil more fertile. Some of the canals are parts of multipurpose projects and, therefore, provide inexpensive source of irrigation. Although the initial cost involved in canal irrigation is more, it is quite cheap in the long run.

Drawbacks of Canal Irrigation: The canal water soaks into the ground and results in water-logging along the canal route. Excessive flow of water in the fields raises the ground water level. Capillary action brings alkaline salts to the surface and makes large areas unfit for agriculture. Huge areas in Panjab, Haryana and Uttar Pradesh suffer from the problem caused by canal irrigation. The muddy areas near the canals act as reproduction grounds of mosquitoes which result in widespread malaria. Many canals overflow during rainy season and flood the surrounding areas. Canal irrigation is suitable in plain areas only.

Surface irrigation structures

Surface irrigation systems are supported by a number of on- and off-farm structures which control and manage the flow and its energy. In order to facilitate efficient surface irrigation, these structures should be easily and cheaply constructed as well as easy to manage and maintain. Each should be standardized for mass production and fabrication in the field by farmers and technicians.

It is not the intent of this guide to be comprehensive with regard to the selection and design of these structures since other sources are available, but it is worthwhile to note some of these structures by way of presenting a larger view of surface irrigation.

The structural elements of a surface system perform several important functions which include: (1) turning the flow to a field on and off; (2) conveying and distributing the flow among fields; (3) water measurement, sediment and debris removal, water level stabilization; and (4) distribution of water onto the field.

Diversion structures

Most surface irrigation systems derive their water supplies from canal systems operated by public or semi-public irrigation departments, districts, or companies. Some irrigation water is supplied in piped delivery systems and some directly pumped from groundwater. Diversion structures perform several tasks including (1) on-off water control which allows the supply agency to allocate its supply and protects the fields below the diversion from untimely flooding; (2) regulation and stabilization of the discharge to the requirements of field channels and watercourse distribution systems; (3) measurement of flow at the turnout in order to establish and protect water entitlements; and (4) protection of downstream structures by controlling sediments and debris as well as dissipating excess kinetic energy in the flow. A typical turnout structure is shown in Figure 5.

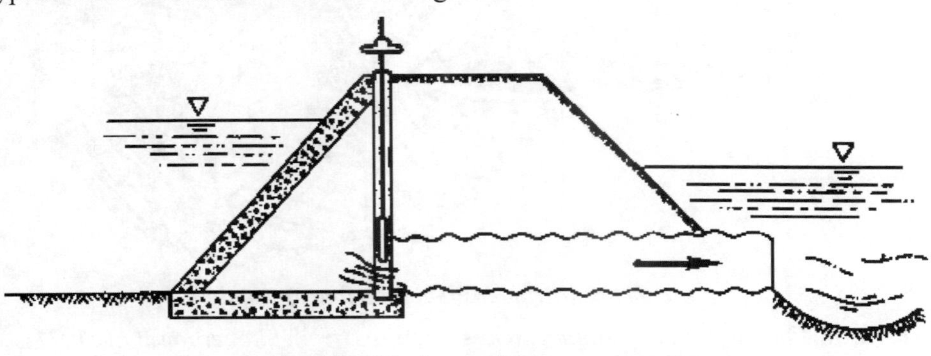

Conveyance, distribution and management structures

Conveying water to the field requires similar structures to those found in major canal networks. The conveyance itself can be an earthen ditch or lateral, a buried pipe, or a lined ditch. Lined sections can be elevated as shown in Figure, or constructed

a. A simple drop structure

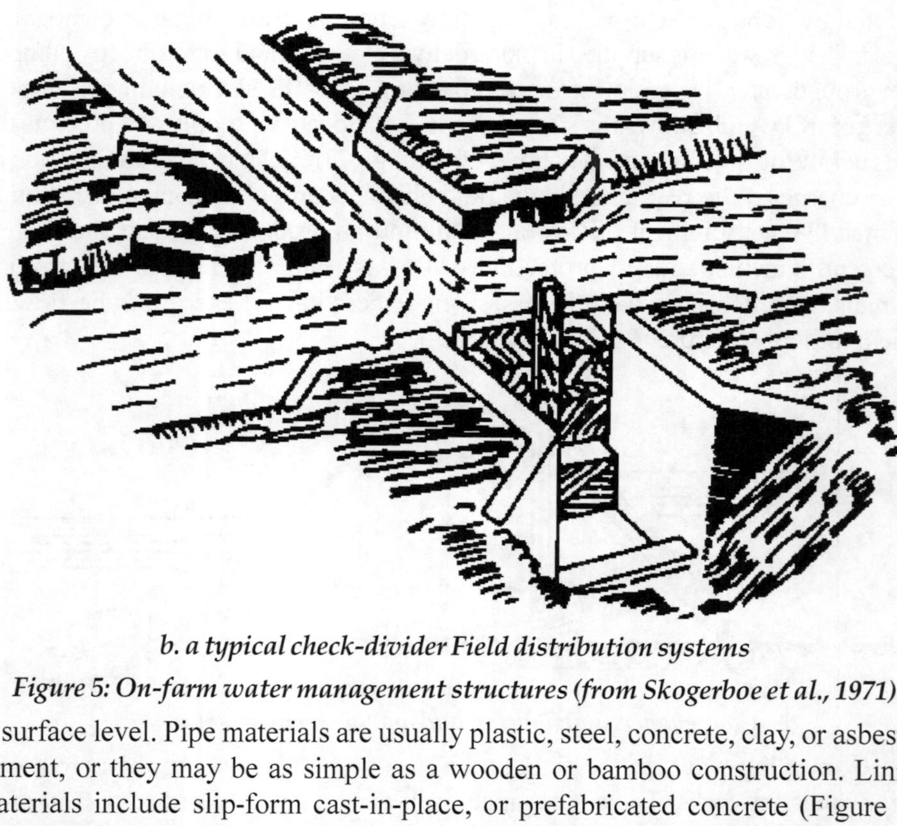

b. a typical check-divider Field distribution systems

Figure 5: On-farm water management structures (from Skogerboe et al., 1971)

at surface level. Pipe materials are usually plastic, steel, concrete, clay, or asbestos cement, or they may be as simple as a wooden or bamboo construction. Lining materials include slip-form cast-in-place, or prefabricated concrete (Figure 9), shotcrete or gunite, asphalt, surface and buried plastic or rubber membranes, and compacted earth.

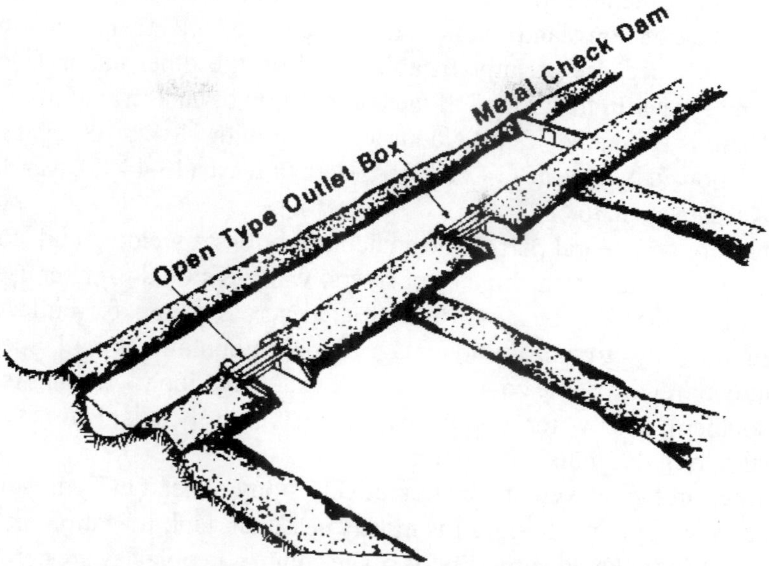

Figure 6: Head ditch outlets for borders and basins (after Kraatz and Mahajan, FAO, 1975)

The management of water in the field channels involves flow measurement, sediment and debris removal, divisions, checks, drop-energy dissipators, and water level regulators. Some of the more common flow control structures for open channels are shown in Figure 6. Associated with these are various flow measuring devices like weirs, flumes, and orifices. The designs of these structures have been standardized since they are small in size and capacity. Designs for flow measurement and drop-energy dissipator structures need more attention and construction must be more precise since their hydraulic responses are quite sensitive to their dimensions.

After the water reaches the field ready to be irrigated, it is distributed onto the field by a variety of means, both simple and elaborately constructed. Most fields have a head ditch or pipeline running along the upper side of the field from which the flow is distributed onto the field.

In a field irrigated from a head ditch, the spreading of water over the field depends somewhat on the method of surface irrigation. For borders and basins, open or piped cutlets as illustrated in Figure are generally used. Furrow systems use outlets which can be directed to each furrow.

Tank Irrigation

The Tank irrigation is more in the rocky plateau area of the county, where the rainfall is uneven and highly seasonal. The Eastern Madhya Pradesh, Chhattisgarh, Orissa, Interiors of Tamil Nadu and some parts of Andhra Pradesh have more land under tank irrigation. It is an old?established practice, particularly in semi?arid tropical parts of South India, but is today a neglected resource, and irrigation tanks are underutilized due to mismanagement.

The development of the means of irrigation like wells and tube wells is not possible because of stony land on one hand: it is easy to collect rain water in natural or artificial pits because of impermeable rocks on the other hand. There is no permeation of water in the ground. If the contribution of tank irrigation in the total irrigated land is estimated, Tamil Nadu ranks first with 18.42% of total irrigated land. It is followed by Odisha (14.60 %), Andhra Pradesh (13.44 %), Kerala (10.26 %) and Karnataka (6.36 %).

Tanks are an integral part of the ancient tradition of yielding and storing the water from rainfall and from streams or rivers. Water is stored for later use mainly in agriculture. Basically tanks are constructed across slopes for collecting and preserving water by utilizing the advantage of local mountains as well. Specifically in the southern region of the country, usage of tank irrigation is critical. As in some parts of southern India, water supply replacement is entirely reliant on a cycle of dry seasons and irregular monsoon seasons.

Further, in the uneven and comparatively stony plateau of Peninsular India, where the range of annual rainfall is mostly seasonal, tank irrigation method has become highly popular. In various parts of the country, people have restored to tank irrigation. In this procedure, a small dam that is constructed across a stream impounds water that is led through slender channels to the cultivated lands. Several parts of states like Madhya Pradesh, Tamil Nadu, Orissa and Andhra Pradeshhave maximum portion of lands under tank irrigation as compared to well or canal irrigation. The main reason behind the high usage of tank irrigation in these states is that canal and well irrigation is possible only in large alluvial river valley plains.

In other parts of the plateau of peninsular India canals and wells are difficult to excavate because of the hard rock below the thin layer of the soil. That is why, in a huge part of the Indian plateau, no other method but tank irrigation is apposite. However, tank irrigation is less economical than other methods of irrigation. There are some major drawbacks of tank irrigation in India and these are mainly - firstly, evaporation of water is a bit rapid because of the large expanse of shallow water of the tanks, secondly, tanks occupy a huge area of the cultivable field and thirdly, tank irrigation do not ensure perennial supply of water. Moreover, at times it also happens that the tanks go dry during the hot summer season when it is urgently required. Thus, by considering the large area that is made useless by the tanks and the great amount of water loss by evaporation, the procedure of tank irrigation is moderately costlier. Advantages of Tank Irrigation: Most of the tanks are natural and not expensive for their construction. Even an individual farmer can have his own tank. Tanks are normally constructed on rocky bed and have long life. In many tanks, fishing is also done. This supplements both the food resources and income of the farmer.

Drawbacks of Tank Irrigation: Major problem with tanks water storage is that tanks dry up during the dry season and fail to provide irrigation when it is needed the most. Silting of the tank bed is a serious problem and it requires de-silting of the tank at regular intervals. Much water is evaporated from the large expanse of shallow water and is therefore not available for irrigation. Tanks cover large areas of cultivable land. In many areas, other sources of irrigation have been adopted and the dry beds of tanks have been reclaimed for agriculture. Furthermore, lifting of water from tanks and carrying it to the fields is a tiring and expensive effort which discourages the use of tanks as a source of irrigation.

MICRO-IRRIGATION OR LOCALIZED IRRIGATION

Drip Irrigation

Drip irrigation is a type of micro-irrigation that has the potential to save water and nutrients by allowing water to drip slowly to the roots of plants, either from above the soil surface or buried below the surface. The goal is to place water directly into the root zone and minimize evaporation. Drip irrigation systems distribute water through a network of valves, pipes, tubing, and emitters. Depending on how well designed, installed, maintained, and operated it is, a drip irrigation system can be more efficient than other types of irrigation systems, such as surface irrigation or sprinkler irrigation.

In the area of irrigation process, drip irrigation is modern technique. It is also called trickle irrigation, which was originally developed in Israel in the early 1960s and became popular in areas of water scarcity. The drip irrigation is the most competent and it can be practised in array of crops, especially in vegetables, orchard crops, flowers and plantation crops (Mamata Swain, 1999). Drip irrigation was used to the ancient custom in certain parts of India of irrigating a tulsi plant kept in the courtyard. During the summer months, the plant was irrigated by a hanging pitcher containing water and a minute hole at its bottom to allow the trickling of water on to the plant. The tribal farmers of Arunachal Pradesh practised a primitive form of drip irrigation system using a slender bamboo as the conduit for water flow. The use of drippers in sub-surface irrigation network was first experimented in Germany in 1869. The noticeable growth of the petrochemical industry during and after the 1950s aided manufacturing of plastic pipes at a cost much cheaper than the cost of metallic or cement concrete pipes. Plastic pipes are convenient for water conveyance under pressure and the plastic material are easily formed into the desired configuration. These features of plastic made the field-scale use of drip irrigation practicable. The drip system was developed for field crops in Israel in the early 1960s and in Australia and North America in the late 1960s. The area under drip irrigation system in the USA is about 1 M ha, followed by India, Spain, and Israel. In India, there has been a tremendous growth in the area under drip irrigation during the last 15 years. At present, around 3.51akh ha area is under drip irrigation with the efforts of the Government of India, while it was only 40 ha in 1960. Maharashtra (94,000 ha), Karnataka (66,000 ha) and Tamil Nadu (55,000 ha) are some of the states where large areas have been brought under drip irrigation. Many crops are irrigated by the drip method in India with the tree crops occupying the maximum percentage of the total area under drip irrigation, followed by vine crops, vegetables, field crops, flowers and other crops.

In drip irrigation, water is applied near the plant root through emitters or drippers, on or below the soil surface. The soil moisture is kept at an optimum level with frequent irrigations. In this method irrigation water is conveyed on the surface in 12 to 16 mm diameter tubing's fed from large feeder pipes. The water is allowed to drip or trickle slowly through the nozzle or orifices at practically zero pressure. In this way the soil in the root-zone of crops is constantly kept wet. Drip irrigation results in a very high water application efficiency of about 90-95 per cent.

The area under drip irrigation system in the USA is about 1 M ha, followed by India and other countries.

Major component of drip irrigation

1. Pump station
2. By pass assembly
3. Control valves.
4. Filtration system
5. Fertilizer tank/venturi
6. Pressure gauge
7. mains/Sub mains
8. Laterals
9. Emitting devices
10. Micro tubes.

Pump station takes water from the source and provides the right pressure for delivery into the pipe system. Control valves control the discharge and pressure in the entire system. Filtration system cleans the water. Common types of filter include screen filters and graded sand filters which remove fine material suspended in the water.

Fertilizer tank/venturi slowly add a measured dose of fertilizer into the water during irrigation. This is one of the major advantages of drip irrigation over other methods.

Mainlines, submains and laterals supply water from the control head into the fields. They are usually made from PVC or polyethylene hose and should be buried below ground because they easily degrade when exposed to direct solar radiation. Lateral pipes are usually 13-32 mm diameter.

Emitters or drippers are devices used to control the discharge of water from the lateral to the plants. They are typically spaced more than 1 metre apart with one or more emitters used for a single plant such as a tree. For row crops more closely spaced emitters may be used to wet a strip of soil. Many different emitter designs have been produced in recent years. The basis of design is to produce an emitter which will provide a specified constant discharge which does not vary much with pressure changes, and does not block easily.

In India, there has been a fabulous growth in the area under drip irrigation during the last many years. At present, major area is under drip irrigation with the help of the Government of India. Reports indicated that Maharashtra (94,000 ha), Karnataka (66,000 ha) and Tamil Nadu (55,000 ha) are some of the states where large areas have been brought under drip irrigation. Many crops are irrigated by the drip method in India with the tree crops occupying the maximum percentage of the total area under drip irrigation, followed by vine crops, vegetables, field crops, flowers and other crops.

The advantages of drip irrigation are under:

1. Possibility of using soluble fertilizers and chemicals.
2. Fertilizer and nutrient loss is minimized due to localized application and reduced leaching.
3. Water application efficiency is high.
4. Field levelling is not necessary. Fields with irregular shapes are easily accommodated.
5. Recycled non-potable water can be safely used.
6. Soil type plays less important role in frequency of irrigation.
7. Soil erosion is lessened.
8. Weed growth is lessened.
9. Water distribution is highly uniform, controlled by output of each nozzle.
10. Labour cost is less than other irrigation methods.
11. Variation in supply can be regulated by regulating the valves and drippers.
12. Plants remains dry, reducing the risk of disease.
13. Usually operated at lower pressure than other types of pressurised irrigation, reducing energy costs.

Sprinkler Irrigation

The sprinkler irrigation system is one of the most efficient irrigation systems that farmers are relying on today. This method is ideal for water conservation since it sprays water evenly through a system of pipes. The entire soil is effectively irrigated through the spray heads. These sprinklers are especially ideal for efficient coverage of small to large areas. The discharge capacity of these sprinklers can be selected according to the irrigation requirement and the type of property.

When choosing a sprinkler irrigation system, you can choose sprinklers that spray water close to the ground since these are considered more water-efficient. The other variant is a sprinkler system that sprays a fine mist of water. Water scarcity is a global concern today, and drip irrigation is the best alternative for efficient water use. This particular irrigation system ensures that little to no water gets easily evaporated in the process.

A well-planned watering system can definitely help avoid over-watering of the crop or plants. Maintenance and management of the irrigation system is essential to save water and money. Let's look at some of the common sprinkler irrigation systems available.

The sprinkler irrigation is categorized according to the functions which are mentioned as under:

1. The main irrigation system
2. The supplementary irrigation system
3. The protective irrigation system

The sprinkler irrigation system is effective for irrigation on uneven lands and on shallow soils. It is also suitable to coarse sandy terrain where the percolation loss is more and where as a consequence, the frequency of irrigation required is more. The sprinkler irrigation system is appropriate in rising and falling land where land shaping is expensive or technically not practicable. The elimination of fertile soil cover by land shaping is not advisable. Sprinkler irrigation system can also be espoused in hilly regions where plantation crops are grown.

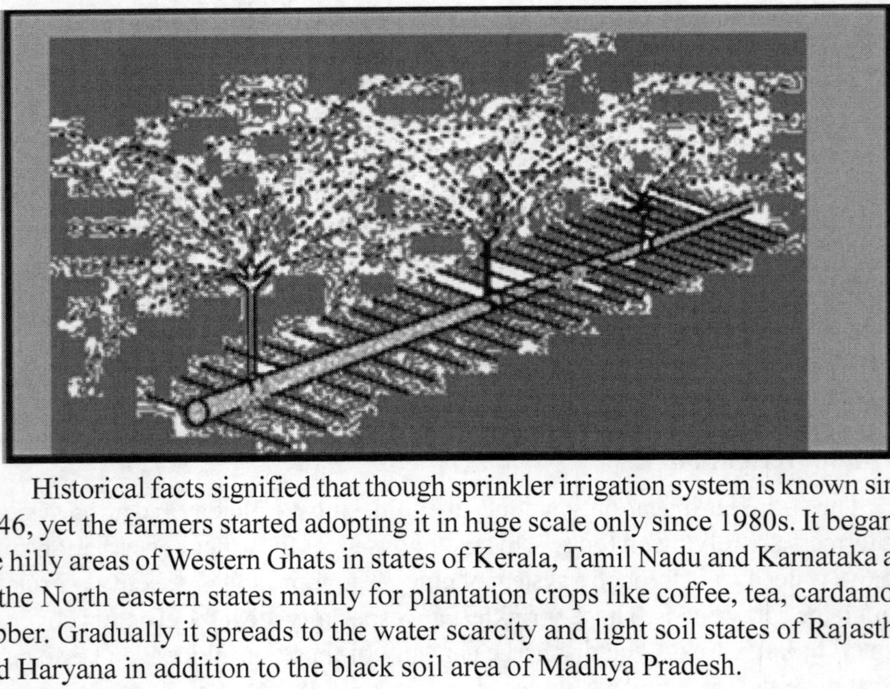

Historical facts signified that though sprinkler irrigation system is known since 1946, yet the farmers started adopting it in huge scale only since 1980s. It began in the hilly areas of Western Ghats in states of Kerala, Tamil Nadu and Karnataka and in the North eastern states mainly for plantation crops like coffee, tea, cardamom, rubber. Gradually it spreads to the water scarcity and light soil states of Rajasthan and Haryana in addition to the black soil area of Madhya Pradesh.

Advantages of sprinkler irrigation:

1. Elimination of the channels for conveyance, therefore no conveyance loss.
2. It is Suitable to all types of soil apart from heavy clay.
3. It is appropriate technique for irrigating crops where the plant population per unit area is very high. It is most suitable for oil seeds and other cereal and vegetable crops.
4. It saves water.
5. With this technique of irrigation, there is control of water application convenient for giving light and frequent irrigation and higher water application efficiency.

6. Sprinkle irrigation increases in yield.
7. There is a mobility of system.
8. It may also be used for undulating area.
9. It saves land as no bunds are required.
10. This technique influences greater conducive micro-climate.
11. Areas located at a higher elevation than the source can be irrigated.
12. In this technique there is a possibility of using soluble fertilizers and chemicals.
13. In this method of irrigation there is less problem of clogging of sprinkler nozzles due to sediment laden water.
14. The overall cost of labour is generally reduced in this method of irrigation.
15. Erosion of soil cover which is common in surface irrigation can be reduced.

Disadvantages of sprinkler system

1. In this technique, initial cost of implementation is high.
2. High and constant energy requirement for operation.
3. Under high wind condition and high temperature distribution and application efficiency is poor.
4. Highly saline water causes leaf burning when temperature is higher than 95 F.
5. When lands have been already levelled and developed for surface or other irrigation methods sprinkler irrigation is not so economical.

There is loss of water due to evaporation from the area during irrigation.

The Factors To Consider

Investing in a sprinkler irrigation system without prior planning or thought can prove to be problematic for you later. There are certain parameters that you must consider before choosing a sprinkler system.

The shape of the field: Before you pick the right sprinkler type and system, you need to consider the shape of your field. If it's an odd shape, you might want to choose the rotor system. If you are seeking even distribution, then opt for either the drip method or overhead sprinkler method, based on your requirement.

The size of the field: The next factor to consider is the size. You cannot ignore the size of your landscape. Based on the size, you have to pick the right sprinkler, and space it strategically at a distance to ensure even distribution of the water. While some crops or plants need slow and calculated distribution of water, others need a steady flow of water to sustain. The time and labour required: Most irrigation systems require adequate maintenance and management. Based on your field and the amount of time and labour required, you can pick the right sprinkler irrigation system. You do not want to invest in an irrigation system unless you are ready to invest the time and effort to maintain it.

The location of the field: The topography of your location is extremely crucial. Whether it's a flat plain or a hilly area, each location requires a specific type of sprinkler head and system for effective irrigation.

Sprinkler System Capacity

The capacity of a sprinkler system is proportional to the flow rate required for irrigation of an area. The capacity is dependent on the crop requirements, the texture of the soil, water-holding capacity of the soil, the root depth and permitted pumping rate as approved by the authorities.

One of the major benefits of investing in a sprinkler system is the low erosion rate. This is why the application rate of the system needs to match the intake rate of the soil. If the system's application is higher than the intake rate, there are chances of water runoff. This will eventually lead to overwatering or underwatering of your field.

Other Irrigation methods

Surge Irrigation: Surge Irrigation is an alternative of furrow irrigation where the water supply is pulsed on and off in planned time periods (e.g. on for 1 hour off for 1½ hour). The wetting and drying cycles reduce infiltration rates resulting in faster advance rates and higher uniformities than constant flow.

Ditch Irrigation: Ditch Irrigation is type of traditional method, where ditches are dug out and seedlings are planted in rows. The plantings are watered by placing canals or furrows in between the rows of plants. Siphon tubes are used to move the water from the main ditch to the canals.

Subirrigation or seepage irrigation: It is a system of irrigation where water is allowed to the plant root zone from below the soil surface and absorbed upwards. The excess may be collected for reuse. Subirrigation is used in growing field crops such as tomatoes, peppers, and sugar cane in areas with high water tables. Major benefits of this system are water and nutrient conservation, and labour-saving. The outfitting cost is comparatively high. Main problems include possibility of increased presence of disease in recycle water.

To summarize, Irrigation is a technique of supplying water to the dry land as a supplementation of rain water. It is mainly aimed for farming. There are various types of systems of irrigation practices in different parts of India. Irrigation in India is carried on through wells, tanks, canals, perennial canal, multi-purpose river valley projects. The irrigation engineer should be acquainted with the type of soil moisture, quality of irrigation water, frequency of irrigation for the proper implementation of irrigation system.

4

Land Drainage Practice

Drains are constructed and are pre-requisite for carrying excess water off agricultural land. Without drains to remove water, the soil becomes waterlogged hampering the growth of crops and pasture. Drainage is essential in establishing good productive land.

There are multiple types of drainage systems. Before designing the drainage system, it is important to consider what will be appropriate for your needs and where the drain is carrying the water to.

Proper planning, design, construction and maintenance of both surface and subsurface drains and channels will minimise the likelihood that they will cause environmental harm and alleviate some of their adverse effects.

It is preferable to drain in the dry months to minimise muddying of downstream waterways. This will also increase the chance that the drains can be revegetated before water starts running again.

Drain Planning

Careful planning should be undertaken when planning drainage.

Prior to drain planning, it is important to consider these points :

- Ensure the land can be drained.
- Determine whether the soil types are able to sustain drainage.
- Check the possibility of acid sulfate soils or salinity occurring.
- Know where the drains will move water to. Will this impact on the downstream environment?
- Know the land capabilities. Will the natural slope of the land sustain drainage? If so, can the drain be changed or modified to suit the application?
- Gain knowledge of the likely flood events or extent of waterlogging. Will your drain have the capacity to pass the size of the flood or amount of water concerned? Find out about Floodplain Mapping for your area.
- Don't over-design the drain.

- Seek appropriate advice according to the prevailing circumstances. This may, in many circumstances, only require some levels to be taken. For larger drainage systems, in-depth soil analysis and design engineering might be required.

EFFECTIVE DRAIN MANAGEMENT

Caring for drains is similar to caring for streams. Activities that occur in the catchments affect the stream once water discharges from the catchment.

Effective drain management is important for preventing environmental problems. Sediments, nutrients (fertilisers), herbicides, pesticides, organic wastes or pollutants washed into the drains flow into the stream and impact on water quality downstream.

Grass cover or other effective plant coverage of the drain should be encouraged to prevent eroding soil moving into the waterway. Grass assists in holding the banks together.

When constructing the drain, it is important to analyse the soil type, as some soils are more prone to erosion than others.

Steep, severe drain banks (batters) should be avoided in most situations. Banks at too steep an angle (batters) are more likely to erode than gentle gradients.

Runoff Speed and Erosion

Drains constructed in sloping country also have the potential to increase water velocity, thus an increase in energy is produced. Energy is caused by increasing the "speed" of the water flow and this has the potential to cause erosive affects, where "head cutting" (gouging out of the bottom and sides of the drain) and drain bank erosion can occur, resulting in sediment transportation and downstream siltation.

There is a tendency among those wishing to drain land to remove floodwaters as quickly as possible from the flood prone areas. Sometimes the "slowing down" of water being drained from an area is one of the most useful and effective methods of minimising the above-mentioned erosive affects. There are several methods that can be considered and applied, such as:

- Retaining areas of bushland, particularly alongside drains, to slow runoff.
- Increase the drain length by not straightening existing waterways, drains or creeks. This reduces the steepness of the drainage system and stops water running off too quickly, thus minimising erosion.
- Construct paddock drains to achieve effective drainage, but on minimal gradients where possible.
- Installation of the smallest but most effective possible outlets from drainage zones. This has the effect of reducing sediment and nutrient transp-ortation by allowing a lesser discharge of water. This method is often used in the Hump and Hollow drainage application, but can be used with the more conventional open ditch applications.
- Installation of simple rock riffles and drop structures. These can be placed in areas where erosion has or is likely to occur. Advice on these structures,

particularly for larger applications, should be sought from technical experts. Engineers such as Department of Primary Industries, Water and Environment River Engineering staff and Regional Water Management Officers are a good first point of contact.

Purpose?

The purpose of the drain must be ascertained so excess water can be moved without damaging the downstream environment. It is possible that removing a flooding problem by building a drainage system in an upstream area could cause, or add to, the severity of flooding downstream. This means the flooding problem has not been rectified, only moved.

Stock

Stock access to the drains should also be controlled, just as stock access to riparian land needs to be managed, to prevent stream bank damage. Fencing drains prevents stock from damaging and depositing organic matter into the drain, and subsequently the catchment. This has the benefit of cutting down on maintenance costs associated with desilting the drain.

Planning of Weeds Control

Weed control methods should be undertaken as a yearly event and should be carefully planned. Failure to do so could result in drains becoming ineffective, and selection of the wrong methods for weed control can be an expensive mistake. If using chemical sprays, selection of the correct chemicals is critical to ensure effective weed control and is critical to ensure animals living in the drain are not killed. Some drains sustain life such as frogs and fish, some of which may be threatened species.

Controlling Weed Growth

There is a need to control weed growth, other foliage growth and siltation in drains to maximise the effectiveness of the drainage system and to reduce further weed spread and flooding occurring.

Drains must be regularly inspected and maintained to achieve long-term effectiveness in performance for removing water.

A typical maintenance program should includes :

- Spraying.
- Removal of weeds.
- Reduction of nutrient sources (that some weeds thrive on!).
- reduction of sunlight by retaining tree growth alongside drains thus minimising weed infestations.
- Inspection to see if stock have broken, and gone through, fences thus damaging drains.
- Land drainage can be distinguished in surface and subsurface drainage.

Principle of Surface Drainage

Surface drainage deals in principle with water-logging on the land surface and subsurface drainage deals in principle with water-logging in the soil, i.e. the water table is too shallow for the purpose the land is used. Interaction between the two kinds of drainage may occur: surface drainage may reduce the need of subsurface drainage and vice versa. To accomplish the drainage function one needs drainage systems, which can be discerned into field and main drainage systems.

Field drainage systems do the actual work of water-logging control while main drainage systems are designed to collect the water from the field drainage systems and transport it to the outlet. The outlet may work by gravity or it may be equipped with a pumping station. Some outlets are designed too work part time by gravity and part time by pumping. Pumping stations may also be found in the interior along main drains. When subsurface drainage is accomplished by wells, these also need a water lifting device.

Field drainage systems can be regular or checked systems. Regular systems work immediately when waterlogging threatens to occur as a result of a considerable recharge, e.g. rainfall, irrigation.Checked systems are systems designed to operate only occasionally.

Examples of application of checked systems can be found in rice growing areas where drainage is normally not required except during certain periods, e.g. during harvest, during exceptionally intensive rainstorms, or to allow the land to dry between crops.

Another example is the closure of drains or the reduction of the drainage capacity during dry spells to conserve water and reduce the losses. Subsurface drainage by wells and pumps constitute automatically checked systems because there is always the possibility to switch off the pumps.

Surface field drainage systems are usually made by land forming/shaping. The regular systems consist of graded land or bedded land using shallow open drains to collect the water. Checked systems consist of bunded basins (in flat land) or bunded terraces (in sloping land). Bunded terraces in sloping land are more often used for erosion control and water conservation than for drainage, yet these sloping lands may occasionally need surface drainage.

Regular subsurface field drainage can be done using subsurface drains viz. mole drains, ditches, tile drains, pipe drains or subsoiling (deep plowing). Mole drains and deep plowing can only be practiced in clay soils.

It is a common misunderstanding that a ditch drain invariably would be a surface drain. Although shallow ditches are used in surface drainage to collect and transport water and may be called surface drains, the deeper ditches can have an important function in the control of the water table and thus are subsurface drains by the definition given above. In fact, subsurface drainage by ditches is more effective than by pipe drains, but sometimes the open ditches may be a hindrance and the maintenance may be a nuisance.

Checked subsurface drainage systems consist of ditches and pipe drains equipped with check structures or pumped outlets as well as of pumped wells.

Drainage Water for Irrigation

If the water quality permits, drainage water can be reused for irrigation. Especially deep wells are useful for this purpose. Thus the water-table may be lowered by pumping during dry spells while an extra underground storage facility is created to be filled up during wet spells. During wet spells it may be desirable to halt or reduce the pumping from wells as no immediate irrigation water is required and the extra storage capacity created in the underground during dry spells now comes in handy.

In irrigated lands of the (semi)arid regions, the drainage water also carries salts and helps to maintain a favorablesalt balance in the soil. All irrigation water, however "sweet", bring salts that remain behind in the soil after evaporation. For example, assuming irrigation water with a low salt concentration of 0.3 g/l (equal to 0.3 kg/m3 corresponding to an electric conductivity of about 0.5 dS/m) and a modest annual supply of irrigation water of 10000 m3/ha (barely 3 mm/day) already brings 3000 kg salt/ha each year. In the absence of sufficient natural drainage (as in waterlogged soils) and without a proper leaching and drainage program, this would lead in the long run to a highsoil salinity that may need to be checked by salinity control.

DRAINAGE IN AGRICULTURAL FIELDS

Principles

Land drainage is principally used to alleviate waterlogging in agricultural fields, but it can also dessiccate certain ground types in prolonged dry spells, so think carefully before installing. Note that land drainage is not the same as a soakaway, although they may both be used to dispose of surface and/or ground water.

Other uses for land drainage are as dispersal drains for septic systems and as collector drains where more formal drainage fittings would be inappropriate. It is also an essential part of civil engineering projects, where the use of fin drains and drainage composites is gradually replacing the more traditional forms of land drainage.

Land drainage works by providing an open conduit for groundwater to follow to a disposal point, or, in the case of dispersal drains, to a dispersal area or leach field. There are a few variations on this theme, involving the type of backfill material, and the type of pre-formed conduit that is used, but all rely on two simple principles; that the land drain provides a 'path of least resistance' for groundwater to follow, and that, left to its own devices, water flows down even the gentlest of slopes.

Groundwater is held by soils within pores, which are the small voids or interstices between the grains that make up the soil. The bedding material surrounding a land drain has significantly larger pores between each stone, providing more space for the groundwater to occupy. The centre of the pipe itself can be thought of as one big pore. Therefore, in a collector drain, there is a 'gradient' of hydrostatic pressure from the soil, via the bedding into the pipe, along which the groundwater will migrate to the point of lowest hydrostatic pressure, ie, the inside of the pipe.

In a dispersal drain, the gradient is reversed because water is being continually added to the pipe from the septic tank or other source, and so the hydrostatic pressure is greatest inside the pipe, causing the flow to be from pipe to soil, rather than from soil to pipe.

Soil mechanics and hydro-dynamics is a major field within the discipline of civil engineering and is far beyond the remit of this site, but, hopefully, this will have given some insight into the basic principles involved in ground water management.

The disposal point of a land drain system, whether it's a manhole, a soakaway, a leach field or an outfall, is always at the lowest point of the system.

Try as we might, we haven't yet managed to get water to flow uphill. During the summer, land drainage can sometimes be identified by lusher growth of vegetation over the line of drainage, and this phenomenon is even more pronounced in situations where land drainage has been used to disperse the processed effluent of a septic tank. This could be turned to your garden's advantage with a little pre-planning.

Why We Need Outfall

Land drainage, whether it has been installed in order to drain an area of soggy ground, or to disperse water collected from hard surfaces or the effluent from a septic tank, must have an outfall, ie, a point where it can 'empty' itself.

Filter and collector systems (see below) often discharge, or "outfall", into a ditch or stream, but they can be connected to the surface water system of a property if there is absolutely no possible alternative, or if your local water board do not permit outfalls to natural watercourses.

Only in the most exceptional of circumstances should land drainage be outfalled to a combined or a foul water system, and, if it is, the connection must be via a trap to prevent smells, effluent and/or raw sewage entering the land drainage system.

For outfalling to a ditch or other watercourse (assuming there is no problem with consent from the land-owner or water authority), the outfall pipe, preferably clay or rigid plastic, needs to be anchored into the bank with a large block of concrete. It's essential that the section of bank directly beneath the outfall pipe is adequately protected from erosion by means of a stone or concrete apron.

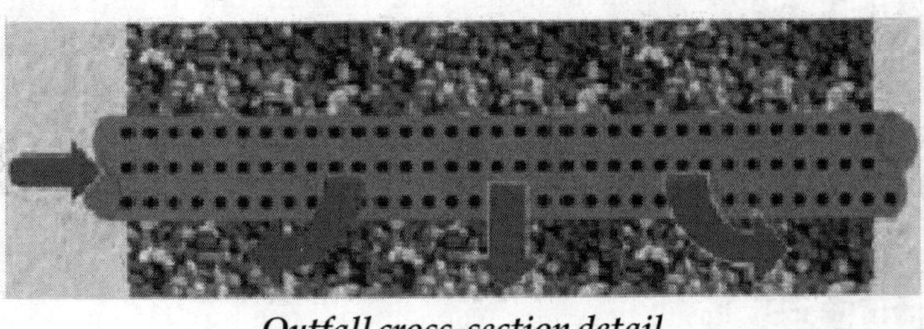

Outfall cross-section detail

Most local authorities and/or water boards have their own specification for an outfall detail, but the illustration opposite shows the the type of scheme favoured by many authorities for small outfalls. Larger outfalls, such as those serving an extensive land-drainage or surface water system, those discharging into a fast-flowing watercourse or those with pipes of diameter greater than 225 mm, may involve cast in-situ concrete structures and possibly steel piling driven into the river-bed. In all cases, consult your local authority before commencing work.

Catch Pits

On filter/collector land drainage systems that have not used a geo-textile to minimise soil particle or sediment migration, a catch-pit should be considered preceding any outfall to a natural watercourse, to remove grits, silts and excess sediment. A catch-pit is, essentially, an empty chamber with an inlet pipe and an outlet pipe set at a level above the floor of the pit. Any sediment carried by the system settles out whilst in the catch pit, from where it can be periodically pumped out or removed.

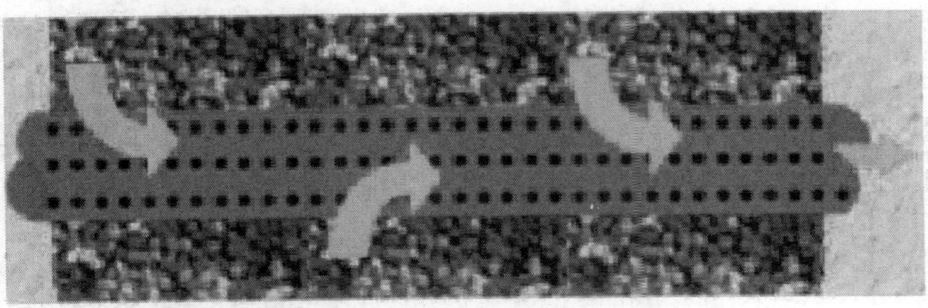

A typical site-constructed catch pit

Manufactured catchpit

Herringbone Pattern

Where land drainage is used to drain a larger garden, it should be installed in the classic herringbone pattern to ensure no point within the area is more than 2.5 m from a drain. Some pre-planning is essential to ensure the best use of the drain and to allow for unavoidable features such as trees, walls, etc., and to ensure that the drainage runs to a convenient outfall at an acceptable gradient. When used to drain fields or other agricultural land, the distance between individual drain runs can be 10 metres or more, depending on local conditions. Such systems are almost always installed by specialist contractors, often on behalf of the local water authority, and a full survey is usually undertaken before the ground is opened. Another use for land drainage in the garden is as a collector drain, installed as a single line, 300-450 mm from the edge of a pavement to carry away the surface run-off and prevent the garden from becoming water-logged. In this sort of scenario, a decorative gravel can be used to dress the surface of the drain, making it a feature of the hard landscaping.

Leach Fields

Dispersal systems, as used with septic tanks, normally discharge into a "Leach Field" (aka an 'absorption field'), a special area downstream of a septic tank where the treated effluent is allowed to soak into the ground whereupon the treatment process is continued by soil-borne bacteria and organisms. The size/area of the leach field is determined by the percolation rate of the ground and the number of persons served by the septic tank. Unlike collector drains, dispersal drains tend to be installed as 'closed' systems, ie, there are no 'dead end' lengths of pipe

Types of Pipes

Most land drainage systems consists of lengths of perforated or slotted plastic or clayware pipe laid in a trench with a porous surround. There is a wide range of sizes, from 80 mm flexible plastic, up to 1000 mm or more for large agricultural or commercial schemes, and they come in a wide variety of materials, including uPVC, clayware, fibre-cement, concrete and ductile iron.

80mm diameter perforated flexible/corrugated plastic pipe is ideal for garden drainage. Larger diameters are also available and they are typically used on larger schemes, such as carriageway edge interceptor drains. The smaller flexible plastic type is readily available in 25m coils from your local builders' merchant at around £1-2 per metre. It is very simple to install, well within the capabilities of a DIY enthusiast. This type of land drain has holes or perforations around the entire circumference, so there is no 'top' or 'bottom'.

Rigid land drain pipes may be fabricated from clayware, plastic or other some other solid material. Some types will be perforated or slotted around the entire circumference while others will be 'half-perforated', ie, have holes on only one face of the pipe. As with their non-perforated cousins, rigid land drain pipes are normally joined together by means of a coupling, although some systems use 'sockets and spigots'.

There are a variety of different types of land drain available. Clayware and rigid plastic perforated pipes are generally more expensive than the flexible type and they should always be used where a land-drain passes beneath a trafficked area, such as a driveway, or where the depth of the drainage is such that the weight of the backfill material would cause a flexible plastic pipe to collapse. As a general rule of thumb, we do not use flexible plastic pipe at depths greater than 1.2 m.

Theoretically, a rigid perforated clayware land drain could be used at depths up to 6m, but this would require machine excavation and shoring, and is best left to professionals.

Land drainage systems rely on a combination of gravity and hydrostatic pressure to create a flow within the pipes. Systems are typically installed with a gentle slope in the region of 1:100-1:200 towards the outfall or disposal point. For this reason, it is best to lay land drainage 'uphill' i.e. start at the outfall, which should be the lowest point on the land drainage system.

The trench width is usually calculated as external pipe diameter (abbreviated to OD, as in Outside Diameter) plus 150 mm sidefill to each side, so, for a pipe with an OD of 115 mm, the trench width would need to be. ..

$$115 + (2 \times 150) = 415 \text{ mm}$$

The trench should be excavated in advance and lined with the geo-textile if one is to be used. The bedding material is placed in the base of the trench, then the pipe itself laid and checked for alignment and gradient before the sidefill is added and finally the backfill.

Ensure a clean gravel is used for the surrounding material, and do not use limestone chippings unless nothing else is available—Limestone can precipitate calcium carbonate in wet conditions, depositing minerals that can clog up the drain. Make sure there is at least 75 mm of gravel around and above the perforated pipe.

Trench sizing

The geo-textile filter is not essential, but will prolong the life of the drain by filtering out smaller clay and soil particles, and deter thirsty roots from seeking refreshment in the land drain. Small rootlets in a land drain can quickly expanded to a pipe-blocking mass!

Popularity of Drainage Trencher

The drainage trencher is probably the most popular machine for installing drainage pipes in all kinds of areas and climates.

The principle allows accurate installation of the pipe and if required backfill material can easily be put over the drainage pipe. Apart from variations in size, there are two distinct types of drainage trenchers. Nowadays, the machine with a steep digging chain design is the most popular around the world but in certain areas the machine with a flat digging chain design has some major advantages in terms of accuracy and productivity.

Use of Drainage Plow

The drainage plow has become the most used machine for installing lateral drainage pipe. The most important reason for this is that it is cheaper than using trenching machines.Production is relatively high but machine running costs are low. This is an important reason and the fact the single blade is the least sensitive for underground obstacles has lead the plow to be the most popular machine for installing lateral drains.

V-plow Vs Single-blade Plow

The V-plow has some very distinct advantages over a single-blade plow and is therefore in some areas the only recognised machine for the trenchless installation of drainage pipe.

In the Netherlands, for example, 95% of all plows are V-plows and the majority of farmers will not accept a single-blade plow in their fields. Other areas that have benefited positively from the use of V-plows include: parts of England, Belgium, some regions of the United States, Mexico, Egypt and India.

Advantages of the V-plow:

- Less deformation or compaction of the soil near the drain, especially when installation is carried out under the so-called "critical depth".
- Less smearing of the trench side.
- Better fissuring which aids water running to the pipe.
- High installation speeds at average depths of less that 5 feet. This advantage increases in clay and clay loam soils.
- Less damage to crops and soil surface.
- Less risk of piping in wet conditions.
- Minimal disruption to the surface so that levelling is not required after installation.
- As a result, use of the V-plow is cheaper than any other installation method.

Collector drainage pipe installation

Collector drainage pipe installation can in principle be done by means of any of the above mentioned machines.

However for larger scale projects normally a large trencher with a steep digging chain design is used with an incorporated cross conveyor. These machines are normally better balanced to compensate for the large trench required and installation process. Mastenbroek has designed machines for some of the largest drainage projects in the world, including specific collector drainage machines for installing corrugated plastic pipes and for installation of concrete pipes by means of the Synchro Trench System.

LEVELLING AND TOPOGRAPHY

Levelling or leveling is a branch of surveying, the object of which is to find the elevation of a given point with respect to the given or assumed Datum. Secondly, it is to establish a point at a given elevation with respect to the given or assumed Datum.

Levelling or leveling is the measurement of geodetic height using an optical levelling instrument and a level staff or rod having a numbered scale. Common levelling instruments include the spirit level, the dumpy level, the digital level, and the laser level.

Spirit levelling employs a spirit level, an instrument consisting of a telescope and a tube level like that used by carpenters, rigidly connected. When the bubble in the tube level is in the middle, the telescope's optical axis (collimation axis) will point exactly in the direction of the local horizontal.

The spirit level is placed on a tripod in the middle between the two points whose height difference is to be determined; the points are marked by markers or benchmarks in the rock or soil. A leveling staff or rod is placed on each point, with measured graduations, usually in centimetres and fractions thereof, or tenths and hundredths of a foot. The observer focuses in turn on each rod and reads the value from it. Subtracting the "back" and "forward" value provides the height difference.

LEVELING PROCEDURE AND INSTRUMENTS

A typical procedure is to set up the instrument within 100 meters (100 yards) of a known or assumed reference elevation point. A rod or staff is held on the point and the instrument is used either manually or automatically to read the rod scale as a backsight. This determines the height of the instrument above the point and allows the height of the instrument (H.I.) above the datum to be computed.

The rod is then held on an unknown point and a foresight reading is taken in the same manner, allowing the elevation of the new point to be computed. The procedure is repeated until the destination point is reached. It is usual practice to perform either a complete loop back to the starting point or else close the traverse on a second point whose elevation is already known. The closure check guards against blunders in the operation, and allows residual error to be distributed in the most likely manner among the stations.

Some instruments provide three crosshairs which allow stadia measurement of the foresight and backsight distances. These also allow use of the average of the three readings (3-wire leveling) as a check against blunders and for averaging out the error of interpolation between marks on the rod scale.

In double-levelling, a surveyor takes two foresights and two backsights and makes sure the difference between the foresights and the difference between the backsights are equal, thereby reducing the amount of error. Double-levelling costs twice as much as single-levelling.

Dumpy Level

The dumpy level is an older style of optical instrument. It is commonly believed that dumpy levelling is less accurate than other types of levelling, but such is not the case. Dumpy levelling requires shorter and therefore more numerous sights, but this fault is compensated by the practice of making foresights and backsights equal.

The Wye level is another older style of instrument, whose operation is similar to that of the dumpy level. Precise Level designs were often used for large leveling projects where utmost accuracy was required. They differ from other levels in having a very precise spirit level tube and a micrometer adjustment to raise or lower the line of sight so that the crosshair can be made to coincide with a line on the rod scale and no interpolation is required.

Automatic Level

Automatic levels make use of a compensator, which ensures that the line of sight is always horizontal regardless of whether or not the housing of the telescope is. This makes it easier and quicker to set-up than a tilting or dumpy level. It also reduces the effect of minor settling of the tripod to the actual amount of motion instead of leveraging the tilt over the sight distance. These instruments became standard in the later part of the twentieth century where three level screws are given to adjust the instrument level.

Digital Level

Digital levels electronically read a bar-coded scale on the staff. These instruments usually include data recording capability. The automation removes the requirement for the operator to read a scale and and write down the value, and so reduces blunders. It may also compute and apply refraction and curvature corrections.

Laser Level

Laser levels project a beam which is visible and/or detectable by a sensor on the leveling rod. This style is widely used in construction work but not for more precise control work. An advantage is that one person can perform the levelling independently, whereas other types require one person at the instrument and one holding the rod.

The sensor can be mounted on earth-moving machinery to allow automated grading.

Topography is the study of Earth's surface shape and features or those of planets, moons, and asteroids. It is also the description of such surface shapes and features (especially their depiction in maps). The topography of an area can also mean the surface shape and features themselves. In a broader sense, topography is concerned with local detail in general, including not only relief but also vegetative and human-made features, and even local history and culture. This meaning is less common in America, where topographic maps with elevation contours have made "topography" synonymous with relief. The older sense of topography as the study of place still has currency in Europe.

Topography specifically involves the recording of relief or terrain, the three-dimensional quality of the surface, and the identification of specific landforms. This is also known as geomorpho-metry. In modern usage, this involves generation of elevation data in electronic form. It is often considered to include the graphic representation of the landform on a map by a variety of techniques, including contour lines, Hypsometric tints, and relief shading.

An objective of topography is to determine the position of any feature or more generally any point in terms of both a horizontal Coordinate system such as latitude, longitude, and altitude. Identifying (naming) features and recognizing typical landform patterns are also part of the field.

A topographic study may be made for a variety of reasons: military planning and geological exploration have been primary motivators to start survey programs, but detailed information about terrain and surface features is essential for the planning and construction of any major civil engineering, public works, or reclamation projects.

The term topography originated in ancient Greece and continued in ancient Rome, as the detailed description of a place. The word comes from the Greek words (topos, place) and (graphia, writing). In classical literature, this refers to writing about a place or places, what is now largely called 'local history'. In Britain and in Europe in general, the word topography is still sometimes used in its original sense.

Detailed military surveys in Britain (beginning in the late eighteenth century) were called Ordnance Surveys, and this term was used into the 20th century as generic for topographic surveys and maps.The earliest scientific surveys in France were called the Cassini maps after the family who produced them over four generations.

The term "topographic surveys" appears to be American in origin. The earliest detailed surveys in the United States were made by the "Topographical Bureau of the Army," formed during the War of 1812. After the work of national mapping was assumed by the U.S. Geological Survey in 1878, the term topographical remained as a general term for detailed surveys and mapping programs, and has been adopted by most other nations as standard.

In the 20th century, the term topography started to be used to describe surface description in other fields where mapping in a broader sense is used, particularly in medical fields such as neurology.

TECHNIQUES OF TOPOGRAPHY

There are a variety of approaches to studying topography. Which method(s) to use depend on the scale and size of the area under study, its accessibility, and the quality of existing surveys.

Direct Survey

Surveying helps determine accurately the terrestrial or three-dimensional space position of points and the distances and angles between them using leveling instruments such as theodolites, dumpy levels and clinometers.

Even though remote sensing has greatly sped up the process of gathering information, and has allowed greater accuracy control over long distances, the direct survey still provides the basic control points and framework for all topographic work, whether manual or GIS-based.

In areas where there has been an extensive direct survey and mapping program (most of Europe and the Continental US, for example), the compiled data forms the basis of basic digital elevation datasets such as USGS DEM data. This data must often be "cleaned" to eliminate discrepancies between surveys, but it still forms a valuable set of information for large-scale analysis.

The original American topographic surveys (or the British "Ordnance" surveys) involved not only recording of relief, but identification of landmark features and vegetative land cover.

Remote sensing is the acquisition of information about an object or phenomenon, without making physical contact with the object. In modern usage, the term generally refers to the use of aerial sensor technologies to detect and classify objects on Earth (both on the surface, and in the atmosphere and oceans) by means of propagated signals.

AERIAL PHOTOGRAPHY

Aerial photography is the taking of photographs of the ground from an elevated position. The term usually refers to images in which the camera is not supported by a ground-based structure. Cameras may be hand held or mounted, and photographs may be taken by a photo-grapher, triggered remotely or triggered automatically. Platforms for aerial photography include fixed-wing aircraft, helicopters, balloons, blimps and dirigibles, rockets, kites, poles, parachutes, vehicle mounted poles . Aerial photography should not be confused with Air-to-Air Photography, when aircraft serve both as a photo platform and subject.

Uses of Imagery

Aerial photography is used in cartography (particularly inphotogrammetric surveys, which are often the basis fortopographic maps), land-use planning, archaeology, movie production, environmental studies, surveillance, commercial advertising, conveyancing and artistic projects.

Radio-controlled Aircraft

Advances in radio controlled models have made it possible for model aircraft to conduct low-altitude aerial photography. This has benefited real-estate advertising, where commercial and residential properties are the photographic subject. Full-size, manned aircraft are prohibited from low flights above populated locations.

Small scale model aircraft offer increased photographic access to these previously restricted areas. Miniature vehicles do not replace full size aircraft, as full size aircraft are capable of longer flight times, higher altitudes, and greater equipment payloads. They are, however, useful in any situation in which a full-scale aircraft would be dangerous to operate. Examples would include the inspection of transformers atop power transmission lines and slow, low-level flight over agricultural fields, both of which can be accomplished by a large-scale radio controlled helicopter. Professional-grade, gyroscopically stabilized camera platforms are available for use under such a model; a large model helicopter with a 26cc gasoline engine can hoist a payload of approximately seven kilograms (15 lbs).

Recent (2006) FAA regulations grounding all commercial RC model flights have been upgraded to require formal FAA certification before permission to fly at any altitude in USA.

Because anything capable of being viewed from a public space is considered outside the realm of privacy in the United States, aerial photography may legally document features and occurrences on private property.

Types of Aerial Photographs

Oblique photographs

Photographs taken at an angle are called oblique photographs. If they are taken from a low angle earth surface–aircraft, they are called low oblique and photographs taken from a high angle are called high or steep oblique.

Vertical photographs

Vertical photographs are taken straight down. They are mainly used in photogrammetry and image interpretation. Pictures that will be used in photogrammetry are traditionally taken with special large format cameras with calibrated and documented geometric properties.

Combinations

Aerial photographs are often combined. Depending on their purpose it can be done in several ways, of which a few are listed below.

- Panoramas can be made by stitching several photographs taken with one hand held camera.
- In pictometry five rigidly mounted cameras provide one vertical and four low oblique pictures that can be used together.
- In some digital cameras for aerial photogrammetry images from several imaging elements, sometimes with separate lenses, are geometrically corrected and combined to one image in the camera.

Orthophotos

Vertical photographs are often used to create orthophotos, photographs which have been geometrically "corrected" so as to be usable as a map. In other words, an orthophoto is a simulation of a photograph taken from an infinite distance, looking straight down from nadir. Perspective must obviously be removed, but variations in terrain should also be corrected for. Multiple geometric transformations are applied to the image, depending on the perspective and terrain corrections required on a particular part of the image.

Orthophotos are commonly used in geographic information systems, such as are used by mapping agencies to create maps. Once the images have been aligned, or 'registered', with known real-world coordinates, they can be widely deployed.

Large sets of orthophotos, typically derived from multiple sources and divided into "tiles" (each typically 256 x 256 pictures in size), are widely used in online map systems such as Google Maps. OpenStreetMap offers the use of similar orthophotos for deriving new map data. Google Earthoverlays orthophotos or satellite imagery onto a digital elevation model to simulate 3D landscapes.

Satellite imagery consists of photographs of Earth or other planets made by means of artificial satellites.

The first images from space were taken on sub-orbital flights. The U.S-launched V-2 flight on October 24, 1946 took one image every 1.5 seconds. With an apogee of 65 miles (105 km), these photos were from five times higher than the previous record, the 13.7 miles (22 km) by the Explorer II balloon mission in 1935.

The first satellite (orbital) photographs of Earth were made on August 14, 1959 by the U.S. Explorer 6. The first satellite photographs of the Moon might have been made on October 6, 1959 by the Soviet satellite Luna 3, on a mission to photograph the far side of the Moon. The Blue Marble photograph was taken from space in 1972, and has become very popular in the media and among the public. Also in 1972 the United States started the Landsat program, the largest program for acquisition of imagery of Earth from space. Landsat 7, the most recent Landsat satellite, was launched in 1999. In 1977, the first real time satellite imagery was acquired by the USA's KH-11 satellite system.

All satellite images produced by NASA are published by Earth Observatory and are freely available to the public. Several other countries have satellite imaging programs, and a collaborative European effort launched the ERS and Envisat satellites carrying various sensors. There are also private companies that provide commercial satellite imagery. In the early 21st century satellite imagery became widely available when affordable, easy to use software with access to satellite imagery databases became offered by several companies and organizations.

Uses

Satellite images have many applications in meteorology, agriculture, geology, forestry,biodiversity conservation, regional planning, education, intelligence and warfare. Images can be in visible colours and in other spectra. There are also elevation maps, usually made by radar imaging. Interpretation and analysis of satellite imagery is conducted using software packages like ERDAS Imagine or ENVI.

Some of the first image enhancement of satellite photos was conducted by the U.S. Government and its contractors. For example ESL Incorporateddeveloped some of the earliest two dimensionalFourier transforms applied to digital image processing to address NASA photos as well asnational security applications. Satellite imagery is also used in seismology and oceanography in deducing changes to land formation, water depth and sea bed, by color caused by earthquakes,volcanoes, and tsunamis.

Resolution and Data

There are four types of resolution when discussing satellite imagery in remote sensing: spatial, spectral, temporal, and radiometric. Campbell (2002) defines these as follows: spatial resolution is defined as the pixel size of an image representing the size of the surface area being measured on the ground, determined by the sensors' instantaneous field of view (IFOV); - spectral resolution is defined by the wavelength interval size (discreet segment of the Electromagnetic Spectrum) and number intervals that the sensor is measuring; temporal resolution is defined by the amount of time (i.e. days) that passes between imagery collection periods; and radiometric resolution is defined as the ability of an imaging system to record many levels of brightness (contrast for example). - Radiometric resolution refers to the effective bit-depth of the sensor (number of greyscale levels) and is typically expressed as 8-bit (0-255), 11-bit (0-2047), 12-bit (0-4095) or 16-bit (0-65,535). - Geometric resolution refers to the satellite sensor's ability to effectively image a portion of the Earth's surface in a single pixel and is typically expressed in terms of Ground Sample Distance, or GSD. GSD is a term containing the overall optical and systemic noise sources and is useful for comparing how well one sensor can "see" an object on the ground within a single pixel. For example, the GSD of Landsat is nearly30m, which means the smallest unit that maps to a single pixel within an image is ~30m x 30m. The latest commercial satellite (GeoEye 1) has a GSD of 0.41 m (effectively 0.5 m due to US Govt restrictions on civilian imaging).

The resolution of satellite images varies depending on the instrument used and the altitude of the satellite's orbit. For example, the Landsat archive offers repeated imagery at 30 meter resolution for the planet, but most of it has not been processed from the raw data. Landsat 7 has an average return period of 16 days. For many smaller areas, images with resolution as high as 41 cm can be available.

Satellite imagery is sometimes supplemented with aerial photography, which has higher resolution, but is more expensive per square meter. Satellite imagery can be combined with vector or raster data in a GIS provided that the imagery has been spatially rectified so that it will properly align with other data sets.

Disadvantages

Because the total area of the land on Earth is so large and because resolution is relatively high, satellite databases are huge and image processing (creating useful images from the raw data) is time-consuming. Depending on the sensor used, weather conditions can affect image quality: for example, it is difficult to obtain images for areas of frequent cloud cover such as mountain-tops.

Commercial satellite companies do not place their imagery into the public domain and do not sell their imagery; instead, one must be licensed to use their imagery. Thus, the ability to legally make derivative products from commercial satellite imagery is minimized.

Privacy concerns have been brought up by some who wish not to have their property shown from above. Google Maps responds to such concerns in their FAQ with the following statement: "We understand your privacy concerns. The images that Google Maps displays are no different from what can be seen by anyone who flies over or drives by a specific geographic location."

Photogrammetry

Photogrammetry is a measurement technique for which the co-ordinates of the points in 3D of an object are determined by the measurements made in two photo-graphic images (or more) taken starting from different positions, usually from different passes of an aerial photography flight. In this technique, the common points are identified on each image. A line of sight (or ray) can be built from the camera location to the point on the object.

It is the intersection of its rays (triangulation) which determines the relative three-dimensional position of the point. Known control points can be used to give these relative positions absolute values. More sophisticated algorithms can exploit other information on the scene known a priori (for example, symmetries in certain cases allowing the rebuilding of three-dimensional co-ordinates starting from one only position of the camera).

Radar and Sonar

Satellite radar mapping is one of the major techniques of generating Digital Elevation Models. Similar techniques are applied in bathymetric surveys using sonar to determine the terrain of the ocean floor. In recent years, LIDAR (Light Detection and Ranging), a remote sensing technique using a laser instead of radio waves, has increasingly been employed for complex mapping needs such as charting canopies and monitoring glaciers.

Forms of Topographic Data

Terrain is commonly modelled either using vector (triangulated irregular network or TIN) or gridded (Raster image) mathematical models. In the most applications in environmental sciences, land surface is represented and modelled using gridded models. In civil engineering and entertainment businesses, the most representations of land surface employ some variant of TIN models. In geostatistics, land surface is commonly modelled as a combination of the two signals—the smooth (spatially correlated) and the rough (noise) signal.

In practice, surveyors first sample heights in an area, then use these to produce a Digital Land Surface Model (also known as a digital elevation model). The DLSM can then be used to visualize terrain, drape remote sensing images, quantify ecological properties of a surface or extract land surface objects. Note that the contour data or any other sampled elevation datasets are not a DLSM.

A DLSM implies that elevation is available continuously at each location in the study area, i.e. that the map represents a complete surface. Digital Land Surface Models should not be confused with Digital Surface Models, which can be surfaces of the canopy, buildings and similar objects.

For example, in the case of surface models produces using the LIDAR technology, one can have several surfaces—starting from the top of the canopy to the actual solid earth. The difference between the two surface models can then be used to derive volumetric measures (height of trees etc.).

Raw Survey Data

Topographic survey information is historically based upon the notes of surveyors. They may derive naming and cultural information from other local sources (for example, boundary delineation may be derived from local cadastral mapping. While of historical interest, these field notes inherently include errors and contradictions that later stages in map production resolve.

Remote Sensing Data

As with field notes, remote sensing data as aerial and satellite photography, is raw and uninterpreted. It may contain holes (due to cloud cover for example or inconsis-tencies (due to the timing of specific image captures). Most modern topographic mapping includes a large component of remotely sensed data in its compilation process.

Topographic Mapping

In its contemporary definition, topographic mapping shows relief. In the United States, USGS topographic maps show relief using contour lines. The USGS calls maps based on topographic surveys, but without contours, "planimetric maps."

These maps show not only the contours, but also any significant streams or other bodies of water, forest cover, built-up areas or individual buildings (depending on scale), and other features and points of interest.

While not officially "topographic" maps, the national surveys of other nations share many of the same features, and so they are often generally called "topographic maps."

Existing topographic survey maps, because of their comprehensive and encyclopedic coverage, form the basis for much derived topographic work. Digital Elevation Models, for example, have often been created not from new remote sensing data but from existing paper topographic maps. Many government and private publishers use the artwork especially the contour lines from existing topographic map sheets as the basis for their own specialized or updated topographic maps.

Topographic mapping should not be confused with Geologic mapping. The latter is concerned with underlying structures and processes to the surface, rather than with identifiable surface features.

LAYING LAND DRAINS

Laying a land drain, or French drain is the easiest method of dealing with surplus water in your garden. Perforated plastic drainage pipe (80mm) can now be bought from many garden centres and all builders merchants in 25 rolls. This type of pipe is easily laid in small trenches and surrounded by shingle (small stones) to allow water through but filter out silt. The pipe can be joined using a waterproof repair tape also available from builder's merchants.

Land Drain Pipe can also be bought in clay sections but the introduction of plastic has made their use, for the domestic application, virtually obsolete. Pipe can also be bought which only has holes half way round. This, for drainage purposes, is laid with the holes uppermost to allow water in from the surface and then to allow it to get carried away completely. We do not advocate the use of this in an English Garden as (See the drought warning below) a fully perforated pipe will allow a more even distribution of the water from wetter, to drier parts, on its journey to the outlet.

With the introduction of what is commonly called Weed Fabric into the market place, it is now possible to wrap the pipe in weed fabric which should stop any silt getting into the pipe. Many builders, ourselves included, actually line the trench with Weed Fabric or Geo Textile Fabric to give it its proper name, to stop any silt getting into the trench. This gives the land drain a much longer life as silt can build up in a matter of weeks and clog all of the voids which are vital to proper drainage.

The Geo Textile Fabric will also stop, or at least dissuade, any local roots seeking out a greater water source. A root system can clog or even break a land drain system quite easily. Geo Textile fabric is a polypropylene mat used to suppress weeds and control soil erosion while allowing the controlled passage of water and air.

Land Drains and French drains can also be used to direct water to a part of your garden which needs more water than others. This is particularly useful for gardeners. A French Drain is simply a land drain but without the pipe although this terminology is getting somewhat mixed these days with many people referring to Land Drains as French Drains.

Land drains in gardens are usually drained to a central point and this is usually a soakaway. A soakaway is simply a hole in the ground filled with rubble and coarse stone with a drainage pipe laid to it removing surface (rain) water from other areas. The soil in which the soakaway is placed must be granular with good drainage properties. It is pointless sinking a soakaway in clay unless there is a more porous layer underneath. The diagram below is not to scale.

A soakaway must be at least 5 metres from any habitable building, by local authority regulations. The pipe flowing to them should be of at least 75mm diameter which is the minimum pipe size for any surface water drainage. 80mm is the recommended size.

This pipe should be laid to a fall of 1 in 100 which means for every 1 m of pipe length the slope should be about 10 mm or 1 cm. However gradual the slope, water will find its way down it.

The size of the soakaway should be a minimum of 1m x 1m square x 1m deep below the bottom of the incoming pipe. The stone infill should surround the pipe and finish approximately 100 smm above it. An impervious layer should then be placed on the stone such as thick polythene, tarpaulin, or even a bed of concrete. Topsoil can then be placed on top of this layer to restore the garden level (A). If you can afford it we also suggest lining the soakaway with weed fabric. As the soakaway needs to be 1m deep below the bottom of the incoming land drain it makes sense to measure the distance of the land drain you intend to lay and work out, using the fall of 1:100, how deep the bottom of the pipe will be when it gets to the soakaway.

For example, if your soakaway is to be placed at the bottom of the garden and your garden is 30m long, the pipe will be 300mm deeper when it gets to the soakaway. If the pipe starts in a trench 300mm deep the bottom of the pipe will be 600mm below the ground when it gets to the soakaway. This means, in total, your soakaway is 1600mm or 1.6m deep.

Care should be taken when digging a hole this deep and we suggest digging the sides back at a fairly shallow angle to avoid any of the sides caving in. Wet soil weighs over 1 tonne per cubic meter and it is extremely dangerous to carry out any excavation over 600mm deep without either shoring up the sides of the trench or raking (digging at an angle) them back. There are no rules to say that a soakaway has to be a perfect cube and your safety is more important than the few extra barrow loads of rubble required to fill a bigger hole.

For turf used in sports, the aim is not merely to remove surplus water, but to remove it quickly enough to allow play as soon as possible. This involves:

- The provision of a drainage system.
- Ensuring the water gets through the soil to the drainage system.

The Benefits of Good Drainage

- Removal of excess water and lowering the water table.
- Improves the quality, firmness and durability of the turf, especially for winter games.

- Quick drying of the soil extending possible playing time.
- Prevention of erosion.
- Prevention of heaving.
- Improved air movement in soils.
- Increased root development.
- Increased capillary moisture in dry weather.
- Improved drought resistance.
- Better soil structure.
- Higher soil temperature and longer growing season.
- Improved bacterial action.

Kinds of Drainage Systems

a. Open drains or ditches.
b. Pipe drains.
c. Mole drains.
d. French drains.

Water on Surface

The cause of the cancelled match or the playing of a match in bad ground conditions is usually because of the water on the surface or in the surface layer of top soil. It takes time for water to penetrate the soil surface and enter drains that are several feet down, so at this point sports field drainage requirements differ from that of the farmer and agriculture, although the basic points are still applicable, such as the laying of pipes, junctions, falls etc.

Types of Drain for Sportsfields

MOLE PLOUGH DRAINING: effective but limited life and the mole cuts must be lined to either a ditch or a drain of some sort - pull mole uphill - clay type soil required: winch pulled mole and/or mole plough fitted direct to tractor on three point linkage. Plough follows ground levels - time of year to carry out work and when not to!

DRAINS - CLAY, POROUS CONCRETE, PERFO-RATED PLASTIC: I prefer to use clay - up to now I have not been impressed with the results of plastic pipes that I have seen laid.

Clay Tiles : For normal work: 3" dia. 4" dia. 6" dia. Plus necessary junction: always use a salt-glazed pipe or similar 3' long at outlet into water course or if connected to storm water system. You have no right of discharge into a surface water sewer. If you get permission to do this you will have to fit a silt trap prior to entry into the sewer, also concrete the face of your outlet into a water course to prevent water undermining it.

Porous Concrete : These drainage pipes can be used except in sulphate-bearing clays such as Gault and London clay. Both contain sulphate and the use of porous concrete drains in these areas is not recommended.

Perforated Plastic : These are now the most widely used pipes for land drainage work. They come in a wide range of sizes - 1 1/2" to 4" - and in lengths of 12 to 20 feet or in large coils. The lengths have a male and female end so that they are easily fitted together. The coils have a sleeve fitting for linking sections together. Because they conduct water more 'smoothly' than clay pipes, a smaller diameter pipe - i.e. 1 1/2" dia. can carry as much drainage water as a 3' clay pipe set at the small rate of fall. This also means that drainage trenches can be narrow, resulting in the removal of backfilling material. They are also lighter to handle and transport than clay and concrete tiles.

NATURAL DRAINAGE

Some soils are well-drained, others easily waterlogged. As rain falls on the turf some evaporates, some may run off and the rest passes into the ground to become soil water. Some of this is absorbed by the plants' roots and is passed back into the atmosphere by transpiration. The rest passes on down until it reaches the level of the water table or an impermeable layer of soil or rock strata. It then commences to form a water table. Water table rises during the winter and falls during the summer. Sandstone is porous and these types or soil are naturally well-drained. Clay-type soils when dry can absorb up to 50% of their bulk in water (like blotting paper), but once this water has been absorbed the clay becomes impermeable and water cannot pass through it.

Excess Water. This means inadequate aeration, lower soil temperature—as much as 12°F degrees between drained and undrained soils. A high water table limits the range of roots. Insufficient air and low temperature slows down the decomposition of organic matter in the soil; and to the groundsman cancelled matches or the destruction of the soil structure if play is allowed to take place, this in turn means difficulty in renovating and establishing new turf during the close season.

Can show drought conditions quicker in the summer. Can these undesirable points be corrected? The reduction of the height of the water table if closer than two feet to the soil surface will enable grass roots to reach their maximum depth. In turn, this improves aeration and mechanical equipment can be used to assist this. Also raises the soil temperature.

The ground and turf are less likely to suffer drought conditions in the summer for it is a fact that in wet weather the driest soil is to be found around the drain, yet in dry weather the soil around the drain has the greatest moisture content.

Drainage Systems

Natural : Where drains follow natural contours, not usually of use in large sports ground work, possibly of use in light soils or on small areas.

Herringbone : Gives short lengths to the lateral drains. More complicated for machine working and the many junctions increase the danger of silting.

Grid System : Easy for machine working. Long laterals. Size of clay drain up to half to half an acre at a fall of 1-300-3". If fall increased the larger areas can be dealt with. After this 4" pipes then use 6" mains unless rate of fall of the main is increased. In laying out a system the aim should be to set the laterals at an angle to the slope so acting as interceptors to the flow of water. Also cut off interceptor drains at banks etc.

Depth of Drains : The outfall - this is usually the governing factor. It should be as deep as possible subject to reason. It should be above the normal level of a watercourse, so as to give unimpeded flow, but deep enough to allow for drains to reach the farthest part of the system at a depth of at least 12" (preferably 18"). In allowing for this, also allow for a fall from lateral drains into main drains. Also, it is generally accepted that the main should fall quicker than the lateral drains to prevent silting and this is also necessary if the same size dia. pipes are used. However, if it is difficult to get sufficient fall then the mains should be of larger size and silt traps built. These can be cleaned out when necessary. There is a rough rule of thumb that for every 1" of depth, the drain will pull for 1 foot on each side: more in light soil, less in heavy. This gives a rough guide to the distance apart of the laterals.

The Fall : This can be as steep as 1-50 if necessary. However, 1-75 is normally regarded as ideal although it can be 1-300, or if dealing with very large areas then 1-600. The important points to bear in mind are that the drain runs are straight and constant in their fall.

The Trench : Nothing is gained in taking out a trench wider than is necessary, i.e. a 4" trench for a 3" pipe. It is a fallacy that more water will enter the pipe by having a trench wider than is necessary to get the pipe into it. The base of the trench should be 'boned' to ensure that there is a constant even fall.

Laying the Pipes : There are two schools of thought:

- Lay the pipes with a slight gap between each pipe not more than 1/8".
- Lay the pipes as close together as possible, i.e. butt the ends tightly to each other.

Back Filling : should be not more than 1 1/2" and smaller if possible. Backfilling should go to within 3-4" of the surface and a binding of fine ash or pea shingle over the top. Then fill with good top soil and either turf or sand.

New Idea : Above the pipe put a 2-3" layer of 3/8" shingle, fill trench with dune sand to within a few inches of the surface, then use light top soil. The idea is that the shingle acts as a valve and once water falls on the surface, so that the dune sand is at field- capacity, then the excess immediately passes from the dune sand into the shingle and then to the drain. The idea is to provide instant drainage yet enable sufficient water to be held in the sand to provide against drought conditions in summer.

The soil around the drain is the driest point in winter and wettest point in summer - hence the dark green colour in the turf.

Laying the Land Drains

To ensure that all applicable parts of your garden are drained, the most effective method of laying land drainsis to lay them in a herring bone pattern. The pipes should be laid so that no point in the garden is any more than 2 m away from a pipe. The maximum distance between pipes is therefore, 4 m. The centre line, or spine, of the drain takes the water to the soakaway and all of the other lines, or ribs, feed into the spine. To set this formation out it is as well to buy a bag of sand and, after measuring the distances involved, sprinkle sand down each line of pipe you want to insert.

Excavation of the trenches should ensure that any turf is cut carefully and stored to be re-laid later. Dig the trenches three times the width of the pipe to the required gentle fall to get the water to the soakaway or outlet.

After lining the trench with weed fabric, lay the pipe in the centre. Surround the pipe with small stone which is available from the builder's merchants and garden centres. Clean stone with an average diameter of 20 mm is fine.

When the pipe is surrounded by an equal thickness of stone, i.e. there should be a top covering of stone equal to, or greater than the side covering, the weed fabric should be folded over the top. At least 150 mm of topsoil is then replaced to lay your turf on. This area can then be rolled to assist with regrowth.

FRENCH DRAINS

A cheaper method of achieving land drainage which is, historically, used to remove surface water from the perimeter of a house or patio, is a French drain.

A French drain is a Land drain without the pipe. It is excavated in the same way, filled with stone in the same way and backfilled in the same way. The construction only differs in so much as it is much more important, with a French Drain, to excavate the bottom of the trench to the exact slope required all the way to the outlet.

With a land drain there is a certain amount of flexibility in so much as the pipe level can be adjusted by some of the bedding.

One further word of warning is required here. The land drains or French Drains you have put in, are there to remove water from areas of your garden. When laid properly they will still be doing this in drier spells and can cause parts of the garden, especially in prolonged dry spells, to scorch. You may notice the lines of the land drains as staying greener and more lush much longer as this is where the water is!

The drainage system was necessary to breathe new life into the property that had been dug up and later refilled. A portion of the Olson Farm in Willmar, Minnesota about 100 miles west of Minneapolis was used to supply fill for a local highway project. A few years later, in late 2006, this soil was replaced by a half a million yards of black dirt from a business park.

"The farm property had a lot of clay underneath, which was taken out to build the four-lane highway, lowering the grade of the property by three feet," explained Vreeman of Vreeman Construction, "so black dirt was brought back to rebuild the field. It took about a year for the new soil to settle."

It was at this point that he started on the drainage system. "Because of all the traffic from the dump trucks packing down the soil both when it was taken out and when the new soil was put in, it was important to get good drainage in there and to get air back into the soil to promote root growth on this property.

"The system will allow any excess water that comes through the soil to drain into the pipe and then off the farm. This keeps the soil moisture at an optimum." Water from the farm moves through the tiling system into a ditch and eventually into the Hawk Creek watershed.

"First, we put in a 12-inch smooth wall pipe in December 2007 that would act as the main," Vreeman explained. "Then, in the spring of 2008 after the soil was drier, we put in the corrugated high-density polyethylene pipe that most call 'seepage tile'. There were a lot of quarter-mile long runs so it went very quickly with the pipe coming directly off the coils that held more than a half-mile of pipe. We put in and connected these runs to the 12-inch main in one day. All 40 acres were planted with corn that same day. We actually started at six in the morning and the planter came in at four o'clock."

The drainage system is set on a grid pattern and consists of the 12-inch main outlet lines, and 24,000 feet of four-inch pipe to collect the seepage. Some eightand six-inch diameter corrugated pipe was also used. The pipe, connections and fittings were manufactured and supplied by Prinsco. The company's environmentally sustainable ECOFLO® corrugated HDPE pipe was used for the 12-inch mainline.

"What's unique about our ECOFLO pipe is that it is manufactured using a blend of material containing a minimum of 50 per cent recycled HDPE," explained Joe Larkins, director of quality for Prinsco. "There is a great deal of technology and scrutiny that goes into our polymer science and manufacturing operations to assure that our customers receive environmentally conscious products with unparalleled performance. Their product is manufactured with recycled materials targeting a 30-hour Notched Constant Ligament Stress (NCLS) value to ensure superior resistance to cracking and a long service life.

Prinsco's GOLDLINE® corrugated HDPE pipe provided the required perforation pattern for the subsurface drainage. The 4-inch diameter pipe was delivered to the farm in 3,000 foot-long coils. Geotextile wrap was not required and installation was done with a self-contained plow, eliminating the need for a more expensive and time consuming cut and cover operation.

The all important grid pattern for the drainage system was mapped-out by Bonnema Surveys.

The most critical factor is a good outlet for the water. Every piece of land lies differently and we strive to lay out the pipe system in such a way that it maximizes the productivity and crop yield of the acreage.

That property was a little different than most because they covered about 80 per cent of the farm with the seepage system. This land has a few low spots where water will stand, but it's fairly flat. And that makes the layout all the more important.

They designed the system to maximize the grade, which leads to a longlife, healthy drainage system. This gives the farmer higher yields, greater crop margins and increases the enjoyment of farming.

Proper soil drainage is one of the most important success factors for a farm from both an environmental and productivity level. It started out nearly 200 years ago with foot-long clay tiles laid by hand to manage the subsurface water. As technology evolved it quickly became apparent that flexible and durable pipe made from high-density polyethylene could be made in coils of several thousand feet. This would not only permit a system to have optimum water flow and maximize crop productivity, but would also be quick to install and last for a very long time.

HDPE pipe is also versatile enough to be manufactured with various perforation patterns. This enables water to drain from the soil no matter what the contour or condition of the field may be and without diminishing the performance or longevity of the pipe.

Surveyor and drainage layout expert Bonnema said "This was a typical project. We went in and took some preliminary elevations to see how to best utilize the natural grade of the farm land to insure a long-lasting functional layout of corrugated polyethylene pipe under the property. Basically, what we're doing is giving the farm land three feet of non-saturated and aerated soil so that the crops will be healthy and productive.

"We've been laying out drainage tile systems for about 30 years. The biggest change is that when we started it was all foot-long concrete tiles laid end to end. Then the corrugated HDPE pipe came along and completely took over the market. It's just a wonderful product. You rarely have a washout."

This was a major problem for farmers, according to Bonnema, especially in Minnesota. "Washouts were a big maintenance headache in the old days. If the frost action or shifting soil caused damage to one of those foot-long concrete tiles, the storm runoff water would eventually dig out a sink hole and a couple of yards of top soil would end up flowing down into the tile system, clogging the tile. And that would have to be dug up and repaired. You won't have that problem with HDPE pipe because it's a continuous pipe. Plus it's a much easier product to ship to the field and install. If it's installed properly it will last many, many years." Known for the coldest winters in the nation, the frostline is also a concern. "Our frost can go down six feet here. That's another thing … HDPE pipe can take the frost where the historic concrete field tile will deteriorate. These days we install HDPE pipe 100 per cent of the time," he said.

Prinsco has been producing drainage solutions for the agricultural, construction and landscaping industries since 1975. Based in Minnesota, it has facilities in the Midwest and the west coast. Prinsco is a PPI member company and part of the Institute's corrugated plastic pipe division.

FIXING THE GRADE OF DRAINS

Every yard lies at a slight, or sharp, angle, and this angle helps drain excess rainwater from your house to the runoff drain. Incorrectly repairing faulty grading that allows rainwater to gather along your foundation where it can erode soil and potentially cause foundation problems can worsen existing problems. Additionally, it can lead water to neighboring yards or cause water to pool. Correctly regrading your yard will alleviate drainage problems and keep your home's foundation sound and solid.

Hammer one stake in the yard that needs grading repair 12 inches away from your home and the other stake at the far end of your yard using the rubber mallet. Hammer them to a depth of 18 inches to keep them solidly fixed within the ground.

Spray paint three "depth marks" on the twine at 1/4, 1/2 and 3/4 intervals using the orange spray paint.

Calculate your "finished depth grade," knowing that for proper rainwater drainage, the grade should descend away from your foundation at a decline of 12 inches every 50 feet. If your yard measures 100 feet long then the end stake should equal 24 inches, the 3/4 mark will measure 18 inches, the 1/2-way mark will measure 12 inches and the 1/4 depth mark will measure 6 inches.

Till the entire yard with the tiller. Tilling loosens the soil, so you can grade it evenly without the tractor's grading blade tearing up large tracts of turf.

Grade the soil in one pass from the house to the first depth mark using the tractor fitted with the grading blade.

Measure the depth between the twine at the 1/4 depth mark and the soil, making sure it equals 25 per cent of the finished depth grade. For instance, if your final depth grade equals 100 feet, the space between the twine and the top of the soil should measure 6 inches because at 100 feet, the final depth grade should equal 24 inches and 6 inches equals 25 per cent of 24.

Regrade the 1/4 depth mark section of the yard as necessary until you reach the desired depth using the tractor fitted with the grading blade.

Grade the 1/2 depth mark section until you reach the desired depth.

Grade the rest of the yard, section by section, until you have graded the yard to the finished depth mark.

Stand 10 yards away from the twine, noticing the smoothness of the decline from the house to the final depth mark.

Rake uneven spots as necessary until the entire yard declines away from the house at a depth of 12 inches per 50 feet.

Reseed the yard with grass seed at a rate of 4 to 6 gallons per 100 square feet. Cover the reseeded yard with straw.

Installing French Drains for Yard Drainage

If your neighbor's land stands at a higher elevation than yours, you may be experiencing problems with excessive moisture on your property. Water from your neighbor's property may be running down the slope and spilling onto your property. You need better yard drainage. One option in such cases is installing French drains.

When some people speak of a "French drain," they refer to a trench in which a drain pipe is laid, but the traditional French drain is basically a trench filled with gravel.

Time Required: Depends on extent of water flow and ground to traverse.

1. Determine a spot on your property where the excess water coming off the slope could be re-routed. Determining such a location may end up being a matter of choosing "the lesser of two evils." If water is currently spilling out at your house foundation and excessive moisture threatens to damage it, obviously almost any other spot would be preferable. The ideal French drain leach field would be an out-of-the-way area with sandy soil, through which the water could percolate harmlessly....
2. But be sure your attempt at yard drainage will not adversely impact anyone else's land.Otherwise, installing a French drain could land you a lawsuit! Check your city codes before digging. Another preliminary step that could save you headaches later is checking with your local utilities concerning the whereabouts of underground cables and the like, so that you'll know where not to excavate for a French drain.
3. Locate the best area for a French drain. Find an area along the slope on your side of the boundary where excavation would be easiest for your French drain (i.e., free of obstructions). Trench lines should be plotted out before you begin digging French drains. You need to create your own mini-slope to carry the water down to its destination. A grade of 1% (i.e., a drop of 1 foot for every 100 feet in length) is often recommended for French drains; others advise a drop of 6" for every 100'. Getting the water to go where you wish is essential for improving yard drainage; the grade will facilitate your efforts.
4. So how do you measure the grading for a French drain? Pound 2 stakes into the ground to mark the beginning and end of the trench. Tie a string tightly to one of the stakes, then run it over to the other stake and tie it off there, too, but loosely (for now). Attach a string level to the string, adjusting the string to get it level. Once it's level, tighten the string at the second stake. Make sure the string is taut. Now begin digging the trench. As you dig, you'll be able to measure down from the string to make sure you are achieving the desired grade for your French drain.
5. Check yourself as you go. For instance, if the trench for the French drain is to be 100' long and the grade 1%, then by the time your trench is 50' long, it should be 6" deeper than where you began excavating.

6. You'll be digging a horizontal trench across the length of the slope. The digging is the most labor-intensive part of installing French drains. The trench will slope down toward the area where you've determined the water will be re-routed (if it doesn't quite reach that spot, you'll have to dig a connecting ditch down to it). Trench width will depend on the magnitude of your moisture problem. Bigger moisture problems call for wider trenches. Small trenches are often dug to a width of 5"-6".
7. Before applying gravel, line the trench with landscape fabric. The landscape fabric will keep dirt out of the gravel. You want to preserve the porosity of the gravel, which promotes percolation of water through it — one of the underlying principles that make covered French drains work. Shovel a coarse gravel onto the landscape fabric. Wrap the ends of the landscape fabric over the top of the gravel layer.
8. You now essentially have a tube of landscape fabric filled with gravel. To fill in the rest of the trench, shovel in a layer of coarse sand, cover it with more landscape fabric, add 4" of topsoil and lay sod on top. Your French drain is complete!

Tips

1. Hire a surveyor. If you don't think you can get the grading right for a French drain on your own, hire a surveyor. Or simply hire a pro to do the whole job.
2. Hire a backhoe operator. If you're not inclined to dig a French drain trench by hand, you could hire a backhoe operator. But that will jack up the cost for the French drain — not only for the digging, but for the extra gravel you'll need (since a backhoe can't dig as small a trench as can a person wielding a spade). Another alternative is suggested by reader, Matt Fisher, noting that "many rental shops rent trenchers now. These machines cut very thin trenches...."
3. Bigger not necessarily better. If you can get away with a small trench for your French drain, you'll save money, as there'll be less gravel needed to fill the trench.
4. Terminology. "Gravel" is a term that can be used differently in different regions. Here, "gravel" refers to small pieces of rock.

Things or Instruments that we needed are :

- spade
- landscape fabric
- gravel
- sand
- sod
- string level
- string
- two stakes
- tape measure

PROPER DRAINAGE

Wet basements, flooded yards, and poor plant growth are symptoms of a poorly drained lot. In Iowa, many homes have been built in poorly drained soils. Wet lots generally are caused by ûooding, springs and seeps, seasonal high water tables, ponding of surface water, or slow soil permeability. This publication explains how to deal with these problems and improve the drainage around your home.

Flooding Houses that are in the ûood plains of a nearby stream or creek may be flooded if the stream overflows during periods of heavy rainfall or rapid snow melt.

To avoid areas that might ûood, check with local ofûcials to determine the flood plain boundaries before buying or building a home.

For existing houses in flood plains, determine how you can ensure the safety of occupants and minimize damage from a flood. Be certain that you have an escape route that will be open during a flood. Use dikes, block outlets that might back up into the basement, waterproof the walls, and block windows and doors with sandbags.

During extreme flooding, water pressure can collapse foundation walls. In some cases it might be preferable to allow the basement to flood and equalize water pressure. Contact local authorities to determine the possibility of basement collapse.

Housing developments often modify the landscape and block or alter natural drainage ways. Check site plans carefully before choosing an existing house or purchasing a lot. You may want to ask a housing professional to help you evaluate the adequacy of the drainage on a particular site. Problems in drainage usually require the cooperation of several homeowners.

Springs and Seeps

Natural springs and seeps occur because of existing soil, rock, and landscape characteristics. Water may flow throughout the year or only seasonally during periods of heavy rainfall.

Water may flow into or around your house if it is constructed over or near a spring or seep. Most Iowa houses require subsurface perimeter drains, at least 4 inches in diameter and surrounded with 6 to 12 inches of gravel or sand and gravel, along the outside of the foundation wall. The bottom of the drain tile must be below the basement ûoor, and the tile must slope to an outlet. When possible, drain perimeter drains to a free outlet, storm sewer, or any approved outlet. If none of these options will work, use a sump pump to remove the water.

Springs and seeps also affect lawns and onsite septic ûelds. Install subsurface drains (tile lines) to remove the water. Tile commonly are made of clay, concrete, perforated plastic, metal, asbestos-cement, or bituminous wood ûber. Check with local building codes for approved materials and other drainage regulations.

Seasonal High Water Table

The water table is the upper surface of groundwater. Below the water table the soil is saturated with water. The level ûuctuates by several feet throughout the year depending on soil, landscape, and weather conditions. In Iowa, the seasonal high water table often is 2 to 5 feet below the ground surface.

In selecting a building site, consider the level of the seasonal high water table. On some sites the seasonal high water table may be at or near the ground surface for long periods. Avoid these areas.

Seasonal high water tables 6 feet deep or more may be of little concern, unless you plan to excavate a deep basement. Subsurface drains such as those used to handle springs and seeps can be used to lower the water table. You'll need a good outlet for the discharge flow and adequate capacity to handle the most severe conditions. Sump pumps should have a back-up power supply in case of power failure.

It is preferable to build the basement ûoor above the seasonal high water table. Provide drainage under the basement ûoor with 4 to 6 inches of free-draining material and a subsurface drainage system. Use a moisture barrier and high quality concrete to reduce water migration through the floor.

Seasonal high water tables around existing houses can be lowered by installing drainage around the outside wall or under the basement floor. Both methods are expensive and disrup tive. Lowering the water table under the basement ûoor should be done with caution. On some soils, especially slow-draining silts and clays, unequal settling may crack the walls.

To remove water from the interior of concrete block walls, install drainage channels at the intersection of the wall and the û oor. These channels are commercially available.

Ponding of Surface Water

You can use small diversions or ditches to channel water away from your lot. Various regulations concerning water flow will apply and you must not change the flow in a way that adversely affects your neighbor. Before beginning a project, check with your neighbors and local ofûcials.

Surface inlets can be used to carry water to a subsurface drain. The drain outlet can empty into street gutters or storm sewers if permitted by local building codes.

Grade your yard so that surface water drains away from the house. Often the wall used around a house settles, leaving a low area. It must be ûlled so water runs away from the house. In the first 10 feet away from the house, the soil should drop a minimum of 6 inches.

A minimum grade of 1 foot in 100 feet is generally adequate after the first 10 feet.

To prevent wood rot, the soil around a house must not be too close to the wood siding. There must be at least 8 inches between the soil and the wood siding. The soil directly around the house should be of slow permeability so water does not penetrate near the house. Do not use rock and gravel on the surface, since they will serve as a conduit for water into the ground.

Avoid using the same subsurface drain for perimeter tile around the foundation and for the downspout. If the capacity of the drain is not sufficient, excess water will be dumped around the basement footings.

Water from downspouts that empty onto the lawn must be dumped and spread far enough from the house so the water does not enter the basement. On steeply sloping well-drained lawns with no basement water problems, a simple splash block usually will be Sufficient. For houses with basement water problems, the water should be dumped at least 5 to 10 feet from the house.

Slow Soil Permeability

Dense layers of soil will restrict the flow of water, and water may pond on the lawn. If the problem is due to poor drainage through the soil, you can drain small wet areas by digging a small trench through the layer and filling it with sand or gravel to improve permeability. However, this method will not work if the problem is caused by a high water table.

Drain larger wet areas by installing subsurface drains 4 to 6 inches in diameter at a depth of 2 to 5 feet. The drains should be back-filled to within a foot of the ground surface with sand or gravel. Use porous topsoil to û ll the last foot.

HOW TO GRADE A LAWN FOR DRAINAGE

Lawn grading around your house should take runoff water away from the foundation. If the soil grading drains water toward your house instead, you can end up with a wet basement or cracks around your foundation.

While most contractors include landscape grading when they finish a new building, older homes often need to have the ground worked so it drains properly. You can do it yourself if you understand the basic principles. In all likelihood, your grass is going to need replanting when you finish the project.

Determine the slope of the land so you know how much lawn grading is necessary. Ideally, you need 1 foot of drop for every 50 feet away from your foundation. Attach one end of a 50-foot length of string to a stake and place the stake in the ground near the foundation of the house. Tie the other end to a second stake and place that stake 50 feet from the house. Use a level to level the string. You may need to do this in several locations around the perimeter of your house.

Use a tape measure to determine how high the string is from the ground at the 50-foot mark. If it's one feet, the slope is just right. If it's less than 12 inches, you should consider adding more slope to your yard. The difference is how much soil you need to grade off the lawn. If it measures more than 12 inches, you have more than adequate slope.

Rent a skid-steer tractor with front loader to move the dirt. Remove 6 to 8 inches of topsoil from the area where you are adding more lawn drainage. Place the topsoil in a pile. Remove the subsoil and put it in a separate pile. Once you remove enough subsoil, replace the topsoil and smooth the surface slope. If you only need to adjust the slope for a small portion of your yard, you may be able to do it by hand with a shovel.

Build up low areas by removing the topsoil. Put the subsoil you removed earlier in the low spots. Cover them with topsoil.

Clean up any rocks, roots or other debris on the ground's surface. Use a rake to smooth the surface of the ground. Flip the rake over and use its back to smooth the surface even more.

MAPS AND RECORDS

A map is a visual representation of an area—a symbolic depiction highlighting relationships between elements of that space such as objects, regions, and themes.

Many maps are static two-dimensional, geometrically accurate (or approximately accurate) representations of three-dimensional space, while others are dynamic or interactive, even three-dimensional. Although most commonly used to depict geography, maps may represent any space, real or imagined, without regard to context or scale; e.g. Brain mapping, DNA mapping, and extra terrestrial mapping.

PRESERVING AGRICULTURAL RECORDS

Why should we be interested in identifying and preserving agricultural records? They are a means by which our culture is passed from one generation to the next, and they provide information from which one generation can learn from its predecessors. Today, there is more interest in family and community history, a need to understand one's "roots". People can learn much about their ancestors, the first immigrants who came to Orange County to reclaim the black dirt from the undrained bogs. Many different ethnic groups were involved with farming in this area.

Agricultural records are indispensable sources for the work of historians and other specialists who interpret the past for present and future use. Many agricultural records have immediate, practical, everyday uses. They are of use to government officials, public interest groups, and to business people. They provide information essential for research into environmental, health and welfare issues, and farming trends.

Agricultural records can be used in the classroom to help students better understand the lives of farmers, the machinery they used, the crops that were planted, the organizations to which they belonged to protect their interests. Younger generations and scholars can use the records to learn about the culture and their ancestors through the use of such records.

By identifying, collecting and preserving farming records now, future generations will have a better understanding of farmers' lives in the present and the past. These records are an important part of Orange County History as well as New York State History.

PRACTICE OF CARTOGRAPHY

Cartography or map-making is the study and practice of crafting representations of the Earth upon a flat surface, and one who makes maps is called a cartographer.

Road maps are perhaps the most widely used maps today, and form a subset of navigational maps, which also include aeronautical and nautical charts, railroad network maps, and hiking and bicycling maps. In terms of quantity, the largest number of drawn map sheets is probably made up by local surveys, carried out by municipalities, utilities, tax assessors, emergency services providers, and other local agencies. Many national surveying projects have been carried out by the military, such as the British Ordnance Survey now a civilian government agency internationally renowned for its comprehensively detailed work.

In addition to location information maps may also be used to portray contour lines (isolines) indicating constant values of elevation, temperature, rainfall, etc.

Orientation of Maps

The orientation of a map is the relationship between the directions on the map and the corresponding compass directions in reality. The word "orient" is derived from Latin oriens, meaning East. In the Middle Ages many maps, including the T and O maps, were drawn with East at the top (meaning that the direction "up" on the map corresponds to East on the compass). Today, the most common – but far from universal – cartographic convention is that North is at the top of a map. Several kinds of maps are often traditionally not oriented with North at the top:

- Maps from non-Western traditions are oriented a variety of ways. Old maps of Edo show the Japanese imperial palace as the "top", but also at the centre, of the map. Labels on the map are oriented in such a way that you cannot read them properly unless you put the imperial palace above your head.
- Medieval European T and O maps such as the Hereford Mappa Mundi were centred on Jerusalem with East at the top. Indeed, prior to the reintroduction of Ptolemy's Geography to Europe around 1400, there was no single convention in the West. Portolan charts, for example, are oriented to the shores they describe.
- Maps of cities bordering a sea are often conventionally oriented with the sea at the top.
- Route and channel maps have traditionally been oriented to the road or waterway they describe.

- Polar maps of the Arctic or Antarctic regions are conventionally centred on the pole; the direction North would be towards or away from the centre of the map, respectively. Typical maps of the Arctic have 0° meridian towards the bottom of the page; maps of the Antarctic have the 0° meridian towards the top of the page.
- Reversed maps, also known as Upside-Down maps or South-Up maps, reverse the "North is up" convention and have South at the top.
- Buckminster Fuller's Dymaxion maps are based on a projection of the Earth's sphere onto anicosahe dron. The resulting triangular pieces may be arranged in any order or orientation.
- Modern digital GIS maps such as ArcMap typically project north at the top of the map, but use math degrees (0 is east, degrees increase counter-clockwise), rather than compass degrees (0 is north, degrees increase clockwise) for orientation of transects. Compass decimal degrees can be converted to math degrees by subtracting them from 450.

Map Types and Projections

Maps of the world or large areas are often either 'political' or 'physical'. The most important purpose of the political map is to show territorial borders; the purpose of the physical is to show features of geography such as mountains, soil type or land use including infrastruction such as roads, railroads and buildings.Topographic maps show elevations and relief with contour linesor shading. Geological maps show not only the physical surface, but characteristics of the underlying rock, fault lines, and subsurface structures.

Maps that depict the surface of the Earth also use a projection, a way of translating the three-dimensional real surface of the geoid to a two-dimensional picture. Perhaps the best-known world-map projection is the Mercator projection, originally designed as a form of nautical chart.

Aeroplane pilots use aeronautical charts based on a Lambert conformal conic projection, in which a cone is laid over the section of the earth to be mapped. The cone intersects the sphere (the earth) at one or two parallels which are chosen as standard lines. This allows the pilots to plot a great-circle route approximation on a flat, two-dimensional chart.

- Azimuthal or Gnomonic map projections are often used in planning air routes due to their ability to represent great circles as straight lines.
- Richard Edes Harrison produced a striking series of maps during and after World War II for Fortune magazine. These used "bird's eye" projections to emphasise globally strategic "fronts" in the air age, pointing out proximities and barriers not apparent on a conventional rectangular projection of the world.

Electronic Maps

From the last quarter of the 20th century, the indispensable tool of the cartographer has been the computer. Much of cartography, especially at the data-gathering survey level, has been subsumed by Geographic Information Systems (GIS). The functionality of maps has been greatly advanced by technology simplifying the superimposition of spatially located variables onto existing geographical maps: Having local information such as rainfall level, distribution of wildlife, or demographic data integrated within the map allows more efficient analysis and better decision making. In the pre-electronic age such superimposition of data led Dr. John Snowto discover the cause of cholera. Today, it is used by agencies of the human kind, as diverse as wildlife conservationists and agricultarists around the world.

Even when GIS is not involved, most cartographers now use a variety of computer graphics programs to generate new maps.

Interactive, computerised maps are commercially available, allowing users to zoom in or zoom out (respectively eaning to increase or decrease the scale), sometimes by replacing one map with another of different scale, centered where possible on the same point. In-car global navigation satellite systems are computerised maps with route-planning and advice facilities which monitor the user's position with the help of satellites. From the computer scientist's point of view, zooming in entails one or a combination of:

1. Replacing the map by a more detailed one
2. Enlarging the same map without enlarging the pixels, hence showing more detail by removing less information compared to the less detailed version
3. Enlarging the same map with the pixels enlarged (replaced by rectangles of pixels); no additional detail is shown, but, depending on the quality of one's vision, possibly more detail can be seen; if a computer display does not show adjacent pixels really separate, but overlapping instead (this does not apply for an LCD, but may apply for a cathode ray tube), then replacing a picture by a rectangle of picture does show more detail. A variation of this method is interpolation.

AGRICULTURAL CROPS OF BLACK DIRT REGION

The black dirt region, originally called "the drowned lands", consists of the remains of a great shallow lake, which formed as the last glaciers were melting away 12,000 years ago. As the ice continued to retreat, the climate warmed and lush vegetation grew, died, and sank to the bottom of the lake. Eventually, most of the lake became swampy, and the decaying organic matter continued to accumulate in the water to a depth in excess of twelve feet in places. The shallow lake lay in this valley from that time until about 100 years ago, when immigrants came to this area and realized the value of the soil which lay below what they then called "the drowned lands."

Orange County produces one half the onions grown in New York State on 5500 acres in the black dirt area. The main type of onion grown is the yellow globe, which is a premium cooking onion. Black dirt onions have been shipped as far away as Israel, and also to the northeastern states and southern state markets. Once known as the "Onion Capital of the United States", the black dirt farmers now face competition from areas having longer growing seasons. Other crops grown in the black dirt area include lettuce, radishes, cabbage, carrots, corn, pumpkin, squash, and there are also sod farms.

Site and Survey Participants

The following list of organizations, farmers, historical societies, and businesses were contacted for the purpose of creating this resource guide. Participants listed in bold print make up this finding aid due to the fact that a substantial amount of both historical and archival material relating to agriculture in the black dirt region were located at those sites. It should be noted, however, that other listed survey participants might also hold similar records but did not participate in the initial or secondary survey.

GRADING THE DITCHES FOR TILE

It is exactly so with the question of the cost of drainage. If the work is insecurely done, and is liable, in five years or in fifty, to become worthless; the increase of the crops resulting from it, must not only cover the yearly interest on the cost, but the yearly depreciation as well. Therefore what may seem at the time of doing the work to be cheapness, is really the greatest extravagance. It is like building a brick wall with clay for mortar.

The bricks and the workmanship cost full price, and the small saving on the mortar will topple the wall over in a few years, while, if well cemented, it would have lasted for centuries. The cutting and filling of the ditches, and the purchase and transportation of the tiles, will cost the same in every case, and these constitute the chief cost; if the proper care in grading, tile-laying and covering, and in making outlets be stingily withheld—saving, perhaps, one-tenth of the expense—what might have been a permanent improvement to the land, may disappear, and the whole outlay be lost in ten years. A saving of ten per cent. in the cost will have lost us the other ninety in a short time.

But, while cheapness is to be shunned, economy is to be sought in every item of the work of draining, and should be studied, by proprietor and engineer, from the first examination of the land, to the throwing of the last shovelful of earth on to the filling of the ditch. There are few operations connected with the cultivation of the soil in which so much may be imperceptibly lost through neglect, and carelessness about little details, as in tile-draining.

In the original levelling of the ground, the adjustment of the lines, the establishing of the most judicious depth and inclination at each point of the drains, the disposition of surface streams during the prosecution of the work, and in the width of the excavation, the line which divides economy and wastefulness is extremely narrow and the most constant vigilance, together with the best judgment and foresight, are needed to avoid unnecessary cost.

In the laying and covering of the tile, on the other hand, it is best to disregard a little slowness and unnecessary care on the part of the workmen, for the sake of the most perfect security of the work.

COSTING AND EXPENDITURE

The items of the work of drainage may be classified as follows:

It is not easy to say what would be the proper charge for this item of the work. In England, the Commissioners under the Drainage Acts of Parliament, and the Boards of Public Works, fix the charge for engineering at $1.25 per acre. That is in a country when the extent of lands undergoing the process of draining is very great, enabling one person to superintend large tracts in the same neighborhood at the same time, and with little or no outlay for travelling expenses. In this country, where the improvement is, thus far, confined to small areas, widely separated; and where there are comparatively few engineers who make a specialty of the work, the charge for services is necessarily much higher, and the amount expended in travelling much greater. In most cases, the proprietor of the land must qualify himself to superintend his own operations, (with the aid of a country surveyor, or a railroad engineer in the necessary instrumental work.) As draining becomes more general, the demand for professional assistance will, without doubt, cause local engineers to turn their attention to the subject, and their services may be more cheaply obtained. At present, it would probably not be prudent to estimate the cost of engineering and superintendence, including the time and skill of the proprietor, at less than $5 per acre, even where from 20 to 50 acres are to be drained at once.

The effect of the silt-basin is to retard the flow of the water, and hence should never be used where this will prove injurious to the drains. The breaking of the current and the entrance friction of the outlet pipe, while it causes the silt to be deposited, also diminishes the velocity and discharge of the main drain, and is detrimental to the action of the drain on level grades.

Where there is abundant fall and the entering drains can discharge into the basin not less than one foot above the top of the outlet drain it may be used.

The best plan for collecting the water of several drains into one is by means of the Y junction and a * 'drop" from branch into main as noted heretofore. The current is preserved, and silt is carried to the outlet before it is deposited. Where the drains are well laid completely graded ditch at the rate of from 10 to 20 rods per hour.

It is governed and operated in the same manner as a traction engine, is compactly built and easily managed by two men. It cannot be expected that a machine of this character will work successfully in ground so soft that it will not bear the weight of the machine, nor in land full of stumps, stones, and large roots. With these exceptions, there seem to be no difficulties which are not successfully met by this machine.

Contracts of Drains

The construction of drains for improving entire farms and large areas of level land is frequently done by contract subject to competent superintendence.

The system is laid out by an engineer and complete plans made for the work, after which bids for the construction of the drains are solicited. The furnishing of the tile and distribution as needed upon the ground is usually not included in the ditching contract, neither is the filling of the trenches.

Tile are purchased by the thousand feet; hauling from the railroad station or factory is done by the ton, the weight of the individual pieces of different sizes being used as a basis for determining the weight of the loads ; the digging of ditches and laying of tile is commonly done by the rod, but by the 100 feet would be better, and filling of the ditches by the 100 feet.

The following general specifications and contract have been found useful in practice, and will serve as a guide in preparing specifications for other work. It will be observed that the only bond or security required of the contractor is the 25 per cent retained until the completion and acceptance of the work. The only method of securing good work of this kind is to give it thorough supervision and know that it is right before the drain is covered.

When a drain has been accepted the liability of the contractor should cease provided the individual line has been finished. The retaining of 25 per cent is made for the purpose of securing the correction of any faults in the work which may be discov- ered before the drain is covered.

SPECIFICATIONS FOR THE CONSTRUCTION OF TILE DRAINS

The lines for the ditches are indicated on the field by stakes which have been set by the engineer, and the depths and grades given by him shall constitute a part of these specifications.

Digging the Ditches. The digging of each ditch must begin at its outlet, or at its junction with another tile drain, arrd proceed toward its upper end. The ditch must be dug along one side of the line of survey stakes, and about ten inches distant from it, in a straight and neat manner, and the top soil thrown on one side of the ditch and the clay on the other.

When a change in the direction of ditch is made, it must be done by means of a neat curve, but in all cases the ditch must be kept near enough to the stakes so that they can be used in grading the bottom. In taking out the last draft, the blade of the spade must not go deeper thr.n the proposed grade line or bed upon which the tiles are to rest.

Grading the Bottom. The ditch must be dug to the depth indicated by the figures given with the survey, which depth is to be measured from the grade stakes which are set for that purpose, and graded evenly on the bottom by means of the " line and gauge ' ' method, or "target," or any other equally accurate device for obtaining an even and true bottom upon which to lay the tile. The bottom must be dressed with the tile hoe, or in case of large tiles, with the shovel, so that a groove will be made to receive the tile, and when laid in it they will remain securely in place.

Laying the Tile. The laying of the tile must begin at the lower end and proceed up-stream. The tile must be laid as closely as practicable, and in lines free from irregular crooks, the pieces being turned about until the upper edge closes, unless there is sand or fine silt which is likely to run into the tile, in which case the lower edge must be laid close, and the upper side covered with clay or other suitable material. When, in making turns, or by reason of irregular-shaped tile, a crack of one fourth inch or more is necessarily left, it must be securely covered with broken pieces of tile. Junctions with branch lines must be carefully and securely made.

Blinding the Tile. After the tile have been laid and inspected by the person in charge of the work, they must be covered with clay to a depth of six inches, unless, in the judgment of the superintendent, the tile are sufficiently firm, so that complete filling of the ditch may be made directly upon the tile. In no case must the tile be covered with sand without other material being first used.

Risk During Construction

The ditch contractor must assume all risks from storms and caving in of ditches, and when each drain is completed it must be free from sand and mud before it will be received and paid for in full. In case it is found impracticable, by reason of bad weather or unlooked-for trouble in digging the ditch, or properly laying the tile, to complete the work at the time specified in the contract, the time may be extended as may be mutually agreed upon by employer and contractor. The contractor shall use all necessary precaution to .secure his work from injury while he is constructing the drain.

Tile to be Used. Tile will be delivered on the ground convenient for the use of the contractor. No tile must be laid which are broken, or soft, or so badly out of shape that they cannot be well laid and make a good and satisfactory drain.

Payments for Work. Unless otherwise hereafter agreed upon, the contractor may at any time claim and receive from the employer seventy-five per cent of the value of completed and accepted work at the price agreed upon in the contract. Twenty-five per cent will be retained until the entire work contracted for is completed and accepted, at which time the whole ount due will be paid.

Prosecution of the Work The work must be pushed as fast as will be onsistent with economy and good workman-ship and must not be left by the contractor for the purpose of working upon other contracts, except by permission and consent of the employer.

Digging Tile Ditches with Machines

Many machines of different patterns have been patented, tried, and have failed to displace hand labor in digging tile ditches. The difficulties to overcome do not seem insurmountable, nevertheless the history of such machines discloses partial successes which for a time promised well, but in the end did not meet the requirements of the work in all kinds and conditions of soils. Ploughs of different patterns have been used to aid the spade by loosening the earth in the bottom of the ditch, thereby diminishing the labor of excavating somewhat.

The machines which have dug the ditch to its full depth at one passage have given the most satisfactory results. The ditch is left completed to grade and ready for the laying of the tile, the excavated earth is left loose and ready to be easily backfilled and the grade is more easily and accurately made than by a machine which excavates by many passages over the ground.

One of these machines which gives excellent results at the present writing and has been more successfully introduced than its predecessors is distinctively a traction ditcher. It is propelled by steam power which operates a cutting wheel and at the same time moves the machine forward by traction, excavating a completely graded ditch at the rate of from 10 to 20 rods per hour. It is governed and operated in the same manner as a traction engine, is compactly built and easily managed by two men. It cannot be expected that a machine of this character will work succcessfully in ground so soft that it will not bear the weight of the machine, nor in land full of stumps, stones, and large roots.

The construction of drains for improving entire farms and large areas of level land is frequently done by contract subject to competent superintendence. The system is laid out by an engineer and complete plans made for the work, after which bids for the construction of the drains are solicited. The furnishing of the tile and distribution as needed upon the ground is usually not included in the ditching contract, neither is the filling of the trenches.

Tile are purchased by the thousand feet; hauling from the railroad station or factory is done by the ton, the weight of the individual pieces of different sizes being used as a basis for determining the weight of the loads ; the digging of ditches and laying of tile is commonly done by the rod, but by the 100 feet would be better, and filling of the ditches by the 100 feet.

The following general specifications and contract have been found useful in practice, and will serve as a guide in preparing specifications for other work. It will be observed that the only bond or security required of the contractor is the 25 per cent retained until the completion and acceptance of the work. The only method of securing good work of this kind is to give it thorough supervision and know that it is right before the drain is covered.

When a drain has been accepted the liability of the contractor should cease provided the individual line has been finished. The retaining of 25 per cent is made for the purpose of securing the correction of any faults in the work which may be discovered before the drain is covered.

Types of Tile

The term “ceramic tile” is a rather loose one, inasmuch as a number of tiling and paving products are included in the description. Ceramic wall tiles made from “slip,” the same material figurines are made from; monocottura (single-fired) floor tiles; rock-hard porcelain tiles; and mosaic tiles made from either porcelain or clay are some of the varieties called “ceramic tiles.”

Wall Tiles

A wall tile is any tile used primarily in applications other than floors. The most prevalent of these are "standard American wall tiles," which also go by the term "four-and-a-quarter," since they are approximately 4 1/4 inches square.

Four-and-a-quarter tiles are those which have been used in bathrooms since the advent of inside plumbing in this country. Many companies make them, and thus there are myriad colors and textures available. Trim pieces or "shapes" are available for every style of standard wall tile made. This is important when it comes to building showers and other structures that require outside corners and edges.

Four-and-a-quarter tiles are almost always formed from "slip," the white ceramic clay used in figurines. The tiles are water-resistant but not water-proof.

There are a number of porcelain tiles which are also designed to be used on walls and surfaces other than floors. Most of these are glazed, and many of them are made in other countries, particularly Asian countries including Japan and Korea. Many times these tiles arrive mounted on backing sheets which facilitate their installation. Porcelain tiles, glazed or unglazed, are virtually water-proof.

Tiles used at the waterlines of swimming pools are usually glazed porcelain tiles. These are almost always imported from other countries, most notably Japan.

Floor Tiles (glazed)

Almost all the glazed floor tiles used today are single-fired "monocottura" tiles. They are made by the process known as "dust pressing," wherein very little water is used, and the bisques or tile bodies are compacted under extremely high pressure.

Glazing is then accomplished, and the tile is cooked at high temperatures, usually over 2000 degrees. A tile produced in this manner is very durable and will last indefinitely.

Grading

In the United States there are five grades assigned to floor tiles, whether they are imported or produced in this country. The grades are in Roman numerals, with grade I being the lowest and V the highest. I know of no one who makes grade I tiles however, so let's start with II.

Grade II tiles can be used in residential applications where light traffic is expected. Bathroom floors, for example, do not receive very heavy traffic.

Grade III tiles are adequate for any residential use, including kitchen and entry floors, which receive considerable traffic. These tiles might also be used in light-traffic commercial applications — perhaps beauty parlors and other businesses where there is not a heavy and continuous volume of foot traffic..

Grade IV tiles are rated commercial, although they are also commonly used in homes. These tiles will hold up in just about any application, including grocery stores, bank lobbies, etc.

Grade V tiles are used in industrial settings, where they are expected to receive heavy abuse and exposure to various chemicals. Not many grade V tiles would be used in the home. They usually look too commercial.

Tile grades are never stamped on packaging, so you are at the mercy of the person selling you the tile. This is one of the reasons we suggest you do your shopping at established tile supply locations and not at discount stores and home centers. Only people who specialize in a product are going to know a great deal about it. Home center employees, for example, simply have too many products to keep up with.

Quarry Tiles

Quarry tiles are made from the same clay bricks are made of. Quarry tiles, including brick pavers, paving bricks and paving tiles are all from the family known as "burnt clay." These tiles come in different sizes and shapes: some are square, some are irregular, and some are generally shaped like bricks laid flat — about four by eight inches or so.

All of the burnt clay materials are extremely durable, inasmuch as they are fired hard, and the color pigments go all the way through from top to bottom. They are rated at grade IV or V. Quarry tiles will never wear out no matter how much abuse they're subjected to.

Porcelain Floor Tiles

Porcelain tiles are probably the hardest products we install. They are dense to the point that they are water-proof, and they are extremely difficult to cut. We wear out more cutting tools on porcelain tiles than on any other product.

Unglazed porcelains are often used in commercial and industrial applications. During manufacturing they can be given non-slip surfaces which make them desirable in outdoor settings and in other instances where the tiles can be expected to become wet. I installed unglazed porcelain tiles in bank lobbies for years — all over Texas, as well as in Louisiana, Arkansas, and Mississippi.

Recently we have encountered more and more glazed porcelain tiles, which are wonderful for residential use. The glaze gives them a warmer look than the unglazed versions. These tiles are generally more expensive than their clay counterparts, though.

Almost all floor tiles used in the U.S. and Canada nowadays are made from porcelain. I don't expect the trend to change. Porcelain tiles have a water absorption rate of .5 or less. There are tiles labeled porcelain that do not meet the absorption standard. If you have any doubt, ask to see the product data. If none are available, move on.

Mosaic Tiles

Mosaics are any tiles, glazed or unglazed, that measure 2 inches square or less, although they don't necessarily have to be square. Some of them, in fact, are hexagonal in shape and others are rectangles. It doesn't matter what the tiles are made of, however most mosaics are made from porcelain, and most of them are left unglazed.

Mosaic tiles are used extensively in commercial applications. Walk into a commercial restroom, and you'll probably notice porcelain mosaic tiles on the floor and possibly on the walls as well.

We often use mosaics to form residential shower floors. The small size makes it feasible to "dish" the shower floor so that water readily flows toward the drain. It is almost impossible to accomplish this with larger pieces and make the job look good. The numerous grout joints between the small tiles provide a great deal of "tooth," and this translates to slip resistance on bare feet.

Mosaic tiles are always mounted on sheets of backing paper or netting — or they are joined together with small spots of synthetic rubber. It has been at least fifty years since they were delivered and installed individually. Can you imagine the time that took?

There are also natural clay tiles, including terracotta tiles and Saltillo tiles. Terracotta tiles are the color of the clay from which they are produced, ranging from grays to browns and reddish orange. They are left unglazed and require a top finish after installation. Saltillo tiles are covered in another article.

An important consideration for all tiles installed outdoors is whether the tiles are "frost-proof." This is especially true in northern climes where freezing conditions are commonplace. Generally, the degree of frost resistance in a tile is determined by its propensity to absorb water. In this sense, tiles that are water-proof are also frost-proof.

Since water expands when it freezes, if it is allowed to get in or under tiles in freezing conditions, it can cause the tiles to crack and come loose. Many ceramic floor tiles, porcelain tiles and all quarry tiles are safe bets for outdoor applications. Terracottas and other natural clay tiles, including Saltillo tiles should not be used in areas where hard freezes occur.

GRADE ITALIAN PORCELAIN TILE

Tile is rated based on a grading system from 1 to 5, with 1 being the lowest or weakest grade of tile and 5 being the highest or strongest and most durable grade. This rating system is used in the United States for both tiles manufactured in the US and tiles imported from other countries. The grade of the tile is not stamped on the tile, so to determine the grade you will need to ask the tile manufacturer or store where the tile is purchased. Italian porcelain tile falls under this grading system as well.

Determine the grade of tile that you need for your project. Grade I tile is typically used for walls because it is so weak. Grade II tile can be used for floors in areas that do not have heavy foot traffic. Grade III tiles are often used in residential buildings for tile floors, counter tops and other areas that do not receive a great amount of use. Grade IV tiles are commonly used in residential buildings as well as offices and commercial buildings and are very durable. Grade V tiles are the strongest tiles and are used in high traffic commercial areas.

Locate a home improvement store or tile store. Shop for and pick out an Italian porcelain tile that will fit your project.

Locate a sales representative or store manager and ask them for the grade of the tile. If the sales representative or manager does not have access to the tile-grading information, write down the name of the tile manufacturer.

Contact the tile manufacturer. Request the manufacturer's contact information from the tile retailer or use an Internet search engine to locate the contact information. For any query clarification, it is advisable to contact the tile grade information from the manufacturing company.

FLOW OF WATER THROUGH PIPES

To understand the operating principles of an electronic drive and motor, it is necessary to understand the basic principles of electricity.

Electricity comes in two forms: alternating current (AC) and direct current (DC). We will first consider the effects of DC on various electrical components and identify the three main characteristics of any electrical circuit.

When we take a close look at nature, we find all matter is composed of atoms. In the basic structure of an atom, we find the nucleus is at the center, surrounded by one or more orbiting electrons. This structure re-occurs millions of times for any material. If the material is an insulator, the orbiting electrons do not move from place to place or from atom to atom.

For the purposes of discussion, we will consider a conductor as an atom with three or less orbiting electrons in the outer shell. An atom with five or more orbiting electrons will be an insulator.

Electrons in the outer orbiting rings find it easy to move from atom to atom (particularly in a conductor like a copper wire) whenever they are forced to do so.

The force that tends to move electrons is voltage (electrical pressure in a circuit). Voltage is basically the force that causes electrons to travel from atom to atom.

As you would expect, the higher the voltage, the more force available to move electrons. Some textbooks use the term electromotive force when describing voltage.

Electrons move from atom to atom to take up a spot vacated by the previous electron. Electrons flow in an orderly manner through a conductor.

Completing the circuit

A typical comparison is to water flow in a pipe with that of electron flow in a conductor. When you turn on a water faucet, a certain amount of water pressure forces water through the pipe and out the end of the faucet.

The exact same phenomena holds true for electricity. When you turn on a light, you allow voltage (force) to push electrons (current) through the wire and cause the light to illuminate.

The obvious question is why is it necessary to move electrons in the first place? The reason is fundamental: Every electrical user (light bulb, TV, toaster, a motor) has resistance, measured in ohms. The user of electricity is an electrical load.

A simple fluid circuit consists of a pump to supply the source of water pressure. The water fountain is the load. The pipes provide the path for the water to flow and also provide a certain amount of resistance to flow.

A simple electrical circuit consists of a source of electrons (battery), a load (light bulb), and conductors (wires) to complete the circuit.

Flow of Water through Pipes Formulas

Velocity of Flow

When water flows through a pipe of uniform cross section, the quantity of water passing any point in a given interval of time depends upon the velocity with which the water flows and the area of cross section of the pipe. It is evident that the quantity of water will equal a column whose cross section is the area of the pipe and whose length is equal to the velocity.

The velocity with which water moves through a pipe is not uniform throughout its cross section. It is least near the wetted perimeter of the pipe where the friction of the pipe retards the flow, and is greatest at the center of the cross section where having to overcome only the friction of its own flowing layers, it attains the maximum velocity. It is assumed in practice, however, that all particles of the water have the same velocity, and the mean of all the velocities in the cross section is taken as the velocity of flow.

Formulas for Velocity

When the size of a pipe and the quantity of water it will discharge in a given time are known, the mean velocity of efflux can be found by the formula: v=q.a In which V = velocity of flow in feet per minute q = quantity of water in cubic feet per minute a = area of cross section of pipe in square feet.*

Example

What must be the velocity of flow in a 2-inch pipe to discharge 6.3 cubic feet of water per minute ?

Solution

Area of pipe = .0213 square feet. Then 6.3/ .0213 - 300 feet per minute.

When the hydrostatic head, length and diameter of a pipe are known, the mean velocity of discharge can be found by the formula:

$$V = m\sqrt{\frac{hd}{1 + 54d}}$$

In which V = mean velocity in feet per second m = coefficient from Table XXVIII d = diameter of pipe in feet h = hydrostatic head in feet 1 = total length of pipe in feet

Table - Values of Coefficient

$\sqrt{\frac{hd}{1+54d}}$	Diameter of Pipe in							
	Feet .05	Feet .10	Feet .50	Feet 1	Feet 1.5	Feet 2	Feet 3	Feet 4
	Ins 5/8	Ins. 11/4	Ins. 6	Ins. 12	Ins. 18	Ins. 24	Ins. 36	Ins. 48
	m	m	m	m	m	m	m	m
.005	29	31	33	35	37	40	44	47
.01	34	35	37	39	42	45	49	53
.02	39	40	42	45	49	52	56	59
.03	41	43	47	50	54	57	60	63
.05	44	47	52	54	56	60	64	67
.10	47	50	54	56	58	62	66	70
.20	48	51	55	58	60	64	67	70

Example

What will be the velocity of discharge from a 6-inch pipe 500 feet long under a head of 60 feet?

*Table of square inches in decimals of a square foot in appendix.

Solution

$$V = 55\sqrt{\frac{60 \times .5}{500 + (54 \times .5)}} = 55\sqrt{\frac{80}{527}} = 55\sqrt{0.567} =$$

$55 \times .298 = 18.09$ feet per second

In column 1 of Table XXVIII will be found that the nearest value corresponding to .238 is .20, and following that line to where it intersects the column headed 6 inches, the value of coefficient m will be found to be 55, which multiplied by the square root .238 gives the velocity sought for. Where great accuracy of calculation is not required, the constant 48, which is an average value of coefficients for small sizes of pipes, can be used, and will give results sufficiently accurate for most practical purposes.

Formula For Head

When the length and diameter of a pipe are known, the head required to discharge a certain quantity of water per second can be found by the formula: h=.000704q21 **d5**

In which h=head in feet l=length of pipe in feet d=diameter of pipe in feet q=quantity of water in cubic feet per second.

Example

What head will be required to discharge from a 4-inch pipe 500 feet long 2 cubic feet of water per second ?

Solution

Substituting values in the formula h=0.000704X4X500 = 333 feet head.
= 0.00422

Formula For Diameter

When the length of a pipe, the hydrostatic head, and the quantity of water required to be delivered per second are. known, the diameter of pipe that will safely take care of that quantity can be found by the formula:

$$d = .234 \sqrt[5]{\frac{q^2 1}{h}}$$

In which d=diameter of pipe in feet q=cubic feet per second to be delivered l=length of pipe in feet h=head in feet Example - What diameter of pipe will be required to deliver .5 cubic foot of water per second through a pipe 2,000 feet long with a head of 400 feet?

Solution

$$-d = .234 \sqrt[5]{\frac{.25 \times 2.000}{400}} = .245 \text{ feet} = 3 \text{ inch pipe}$$

Formulas For Quantity

When the mean velocity and the area of a pipe are known, the quantity of water discharged in a given interval of time can be determined by the formula: q=va In which q=quantity of water in cubic feet per minute v=velocity of flow in feet per minute a=area of cross section of pipe in square feet

Example

How many cubic feet of water will be discharged per minute by a 2-inch pipe when the velocity of efflux is 300 feet per minute?

Solution

Area of 2-inch pipe=.021 square feet. Then .021x300 =6.3 cubic feet. Answer.

When the diameter, head and length of a pipe are known, the quantity of water it will deliver in a given time can be found by the formula:

$$q = \sqrt{\frac{d^2 h}{1}} \times 4.71$$

In which q=quantity in cubic feet per minute d=diameter of pipe in inches h=head in feet l=length of pipe in feet

Example

What quantity of water can be delivered per minute through a 3-inch pipe 2,000 feet long with a head of 400 feet?

Solution

$$-q = \sqrt{\frac{243 \times 400}{2,000}} \times 4.71 = 32.97 \text{ cubic feet per minute}$$

The velocity of flow in drains or pipes running partly full can be found by the formula:

$$V = \sqrt{\frac{a}{p} 2d}$$

In which V=velocity in feet per second a=area of water in square feet P=wetted perimeter in feet 2d=twice the slope in feet per mile

Example

What is the velocity of flow in a 6-inch drain laid at a grade of 1/4 inch per foot when running half full?

Solution

There is a 110-foot fall in a mile of drain laid at a grade of 1/4 inch per foot.

$$V = \sqrt{\frac{.098}{.75} \times 220} = 5.8 \text{ feet per second}$$

This water flow rate table shows water flow through pipes and copper tubing. Volume of flow is at one foot per minute velocity in Pipe or Tube.

Schedule 40 pipe

Dia. inches	Cubic ft/min	Gallons/minute
1/8	0.0004	0.003
1/4	0.0007	0.005
3/8	0.0013	0.010
1/2	0.0021	0.016
3/4	0.0037	0.028
1	0.0062	0.046
1-1/4	0.0104	0.078
1-1/2	0.0141	0.106
2	0.0233	0.174

The Armfield C6-MKII-10 Fluid Friction Apparatus is designed to allow the detailed study of the fluid friction head losses which occur when an incompressible fluid flows through pipes, bends, valves and pipe flow metering devices.

Friction head losses in straight pipes of different sizes can be investigated over a range of Reynolds' numbers from 103 to nearly 105, thereby covering the laminar, transitional and turbulent flow regimes in smooth pipes. A further test pipe is artificially roughened and, at the higher Reynolds' numbers, shows a clear departure from typical smooth bore pipe characteristics.

In addition to the smooth and roughened pipes, a wide range of pipeline components are fitted, including pipe fittings and control valves, allowing investigation of the losses caused by this type of connection. A clear acrylic section of pipeline houses a Venturi meter, an orifice plate assembly and a Pitot tube, so that these can be investigated as flow measurement devices.

The C6-MKII-10 is designed to be operated in conjunction with the Armfield F1-10 Hydraulics Bench. The unit can be used with a range of instrumentation packages including water and mercury manometers, hand-held digital pressure meters and a computer data logging pack.

DESCRIPTION OF EQUIPMENT

ARMFIELD C6-MKII-10 Fluid Friction Apparatus

All numerical references relate to the diagram below

The test pipes and fittings are mounted on a tubular frame carried on castors. Water is fed in from the hydraulics bench via the barbed connector (1), flows through the network of pipes and fittings, and is fed back into the volumetric tank via the exit tube (23). The pipes are arranged to provide facilities for testing the following:

An in-line strainer (2)
An artificially roughened pipe (7)
Smooth bore pipes of 4 different diameters (8), (9), (10) and (11)
A long radius 90° bend (6)
A short radius 90° bend (15)
A 45° "Y" (4)
A 45° elbow (5)
A 90° "T" (13)
A 90° mitre (14)
A 90° elbow (22)
A sudden contraction (3)
A sudden enlargement (16)
A pipe section made of clear acrylic with a Pitot static tube (17)
A Venturi meter made of clear acrylic (18)
An orifice meter made of clear acrylic (19)
A ball valve (12)
A globe valve (20)
A gate valve (21)

Short samples of each size test pipe (24) are provided loose so that you can measure the exact diameter and determine the nature of the internal finish. The ratio of the diameter of the pipe to the distance of the pressure tappings from the ends of each pipe has been selected to minimise end and entry effects. A system of isolating valves (25) is provided whereby the pipe to be tested can be selected without disconnecting or draining the system. The arrangement also allows tests to be conducted on parallel pipe configurations. Each pressure tapping is fitted with a quick connection facility. Probe attachments with an adequate quantity of translucent polythene tubing are provided, so that any pair of pressure tappings can be rapidly connected to the pressure measurement system.

Controlling the Flow

Water is pumped through the Fluid Friction Apparatus using a centrifugal pump mounted on the inside of the hydraulics bench. The pump can be switched on and off using the switch indicated in the diagram shown in the equipment section of this web site. The Control Valve should always be closed before starting the pump.

Water flows through the connector in the channel on the bench top, through the flexible connecting hose shown in the diagram above and into the C6-MKII-10. It will then flow through whichever of the test pipes is selected, back through the acrylic pipe section and into the volumetric tank in the hydraulics bench.

Flow rates through the apparatus may be adjusted by operation of the Control Valve on the hydraulics bench. Simultaneous operation of the flow control valve with the two outlet valves (gate and globe) shown in the diagram above, will permit adjustment of the static pressure in the apparatus together with the flow rate. Using the three valves in combination, it should be possible to achieve fine adjustment of the flow if required.

Flow Path

The flow path through the pipe friction network is controlled using the system of isolating valves shown in the diagram above. By opening and closing these valves as appropriate, it is possible to select flow through any combination of pipes.

The service module incorporates a moulded volumetric measuring tank which is stepped to accommodate low or high flow rates. A stilling baffle is incorporated to reduce turbulence.

A remote sight gauge, consisting of a sight tube and scale, is connected to a tapping in the base of the tank and gives an instantaneous indication of water level. The scale is divided into two zones corresponding to the volume above and below the step in the tank.

A dump valve in the base of the volumetric tank is operated by a remote actuator. In operation, the volumetric tank is emptied by lifting the dump valve, allowing the entrained water to return to the sump. When test conditions have stabilised, the dump valve is lowered, retaining the water in the tank.

Timings are taken as the water level rises in the tank. Low flow rates are monitored on the lower portion of the scale corresponding to the small volume beneath the step. Larger flow rates are monitored on the upper scale corresponding to the main tank.

When very small volumetric flow rates are to be measured, the measuring cylinder supplied with the F1-10 should be used rather than the volumetric tank. When using the measuring cylinder, diversion of the flow to and from the cylinder should be synchronised as closely as possible with the starting and stopping of a watch; do not attempt to use a definite time or a definite volume.

Tapping Points

The head loss due to pipe friction is measured by taking pressure readings at different tapping points on the pipe network. In order to measure the pressure loss along a pipe, the pressure measurement device is connected between a pair of tappings, using the tubing and connectors.

Each pressure point on the apparatus is fitted with a self-sealing connection. To connect a test probe to a pressure point, simply push the tip of the test probe into the pressure point until it latches. To disconnect a test probe from a pressure point, press the metal clip of the side of the pressure point to release the test probe. Both test probe and pressure point will seal to sprevent loss of water.

Operation with Manometers

Flexible tubes are connected to the inlets at the bottom of the manometer and the quick release connectors are fitted to the other ends of the tubes. Connect the manometer tubes to the pipe network at two tappings with a high pressure drop (e.g. either side of a partially closed valve) and start the pump. Water will be forced through the manometer, expelling the air in the pipes.

When all air bubbles have been expelled, disconnect the manometer from the pipe network. The quick release fittings will seal keeping the tubes full of water.

The pressurised water manometer incorporates a Schrader valve which is connected to the top manifold. This permits the levels in the limbs to be adjusted for measurement of small differential pressures at various static pressures.

The hand pump will be required to effect reduction of levels at high static pressures. Alternatively the foot pump may be used.

Operation with Hand Held Pressure Meter

Fit the quick release fittings supplied with the C6-MKII-10 to the ends of the tubes on the hand held pressure meter.

It is important to expel any air which may be trapped in the pipes of the pressure meter before taking readings. Connect the meter tubes to a convenient pair of tappings and switch on the pump. Carefully undo one of the nuts holding the tubing to the pressure meter until liquid is expelled from the joint. Bleed the tube to expel any air. Tighten the nut and repeat for the other tube.

When taking readings with the hand held pressure meter, it is important that the meter is zeroed before taking a set of results. Switch on the pump, close the outlet valves and then close the control valve to leave the system at a high static pressure. When the reading on the meter has stabilised, press the zero button to reset the meter.

Specifications

Test Pipe Diameters:

1. 19.1mm x 17.2mm
2. 12.7mm x 10.9mm
3. 9.5mm x 7.7mm
4. 6.4mm x 4.5mm
5. 19.1mmx 15.2mm (artificially roughened)

Distance between tappings: 1.00m

FLUID FRICTION IN A SMOOTH BORE PIPE

Objective

To determine the relationship between head loss due to fluid friction and velocity for flow of water through smooth bore pipes and to confirm the head loss friction factor.

Method

To obtain a series of readings of head loss at different flow rates ,through one or more of the smooth bore test pipes.

Theory

Professor Osborne Reynolds demonstrated that two types of flow may exist in a pipe.

1. Laminar flow at low velocities where $h \alpha u$
2. Turbulent flow at higher velocities where $h \alpha un$

Where h is the head loss due to friction and u is the fluid velocity, sometimes shown as v in some text books. These two types of flow are separated by a transition phase where no definite relationship between h and u exists.

Graphs of h versus u and log h versus log u show these zones.

Furthermore, for a circular pipe flowing full, the head loss due to friction may be calculated from the formula:

$$h = \frac{4fLu^2}{2gd} \quad \text{or} \quad h = \frac{\lambda Lu^2}{2gd} \qquad \text{(Eq. 1)}$$

(Note that some text books may show v2 instead of u2, but they are both the same thing.)

L is the length of the pipe between tappings,

d is the internal diameter of the pipe,

u , is the mean velocity of water through the pipe in m/s,

g is the acceleration due to gravity in m/s2 and

f is the pipe friction coefficient.

Note that the American equivalent of the British term f is λ where $\lambda = 4f$.

Reynolds' number, Re, can be found using the following equation:

$$Re = \frac{\rho u d}{\mu} \qquad (\text{Eq. } 2)$$

where μ is the dynamic viscosity (1.15 x 10-3 Ns/m2 at 15°C) and ρ is the density (999 kg/m3 at 15°C).

Having established the value of Reynolds' number for flow in the pipe, the value of f may be determined using a Moody diagram, a simplified version of which is shown below. You may have a more detailed version in your notes. There is also a detailed version in "Chris's web site". You will find the link in my useful links section.

Equation 1 can the be used to determine the theoretical head loss if you know the value of f for the pipe. Alternatively you can rearrange the same equation to determine f for the pipe being used in your experiment.

Equipment Set Up

Additional equipment required: Stop watch, Internal Digital Calliper.

Arrange the valves on the C6-MKII-10 to allow flow through only the test pipe under observation.

Procedure

Prime the pipe network with water. Open and close the appropriate valves to obtain flow of water through the required test pipe.

Take readings at a number of different flow rates, altering the flow using the control valve on the apparatus, (ten readings is sufficient to produce a good head-flow curve). Measure flow rates using the volumetric tank. For small flow rates use the measuring cylinder. Measure head loss between the tappings using the portable pressure meter or pressurised water manometer as appropriate.

Obtain readings on all four smooth test pipes if you have the time.

Measure the internal diameter of each test pipe sample using a Digital calliper.

SIZE OF LATERAL DRAINS

Although most drainage is applied to level land, increasingly more drainage systems are being used on sloping lands. However, sloping lands pose special problems. Unlike drainage of level lands, flow velocities and pressure can build up in drainage systems on sloping land. This can pose substantial hazards to the drainage system and to the land.

The purpose of this chapter is to direct attention to potential hazards and to measures of safeguarding against such hazards.

ASSESSMENT OF SLOPE CONDITIONS

It is imperative that all aspects of the drainage system are of the highest quality. A good plan is essential.

This permits accurate assessment of slope conditions and allows for the pinpointing of potential failure points. Plans are important, even when doing small scale installations with a backhoe.

While under-design of drain capacity on level land is of little consequence it can be disastrous on sloping land.

Severe rainstorms can temporarily overload the drains and cause the water to back up resulting in a buildup of water pressure. This rise in water pressure may force water to the surface through perforations or other openings in the line creating the potential for serious erosion.

For these reasons, main lines must never be constructed from perforated pipe. It is essential that the main lines be of adequate size, particularly at the lower end of the system, so as to comfortably accommodate heavy discharge. Connections must be fastened securely and be installed with carefully placed bedding or backfill material to hold the drain line in place.

Private Sewer

A private sewer is a drainage system serving two or more properties, which has not been adopted by the Water and Sewerage Company. A lateral drain is the section of drainage system serving a single property, which lies outside the boundary of that property.

Private sewers and lateral drains are currently the responsibility of the owners and occupiers of the properties they serve. Unless a problem occurs most householders are often unaware that they are responsible for their private sewer (sometimes jointly with others). Even where they are aware of their responsibilities, when problems do occur the costs associated with maintaining and repairing private sewers can be extremely high and are sometimes met by individuals or spread over just a few households resulting in considerable financial burdens for those responsible for them. This disparate ownership, together with a lack of planned maintenance, also means that society does not gain the benefit that integrated management of the sewerage system as a whole would bring, which is increasingly seen as important in ensuring the functionality of the system.

The UK and Welsh Assembly Government consulted on a 'Review of Existing Private Sewers and Drains in England and Wales' following which the UK and Welsh Assembly Governments concluded that transfer provided the only comprehensive solution to the problems associated with existing private sewers and lateral drains in 2003.

The UK Government and Welsh Assembly Government announced an 'in principle' decision that existing private sewers and lateral drains in England and

Wales, connecting to the public sewerage network, should be transferred into the ownership of Water and Sewerage Companies on 22 Februar, 2007.

Following a joint consultation in 2007 on 'Private Sewers Transfer – Implementation Options' on possible methods for implementing the transfer of private sewer ownership, in March 2008, Defra and the Welsh Assembly Government it was clear that the majority of respondents favoured an overnight automatic transfer of ownership of private sewers and lateral drains connected to a public sewer to the statutory Water and Sewerage Companies, though response from small drainage businesses did not.

In the Summary of Response document, Defra and the Welsh Assembly Government made a commitment to draw up recommendations for implementing the transfer of private sewers and lateral drains.

UK Government announced on 15 December 2008 that from 2011 all private sewers and lateral drains that drain to the public sewerage network of water and sewerage companies in England will transfer into their ownership.

The Welsh Assembly Government's Strategic Policy Position Statement on Water, which was published on 31 March 2009, made a commitment to pursue the development of regulations in 2011 to facilitate the transfer of private sewers to water and sewerage undertakers.

A joint consultation on draft regulations and proposals for schemes for the transfer of private sewers and lateral drains to water and sewerage companies ran from August to November 2010.

The minister for environment, sustainability and housing has agreed to the publication of the government response and summary of responses to the consultation on draft regulation and proposals for schemes for the transfer of private sewers and lateral drains to water and sewerage companies in england and wales.

Water and Sewerage Undertakers can, on their initiative, or on the application of the sewer's owner, adopt existing private sewerage assets under section 102 of the Water Industry Act 1991. Section 105A of the Water Industry Act 1991, allows for Welsh Ministers, in relation to water or sewerage undertakers whose areas are wholly or mainly in Wales, to introduce schemes under which those undertakers are required to use their existing section 102 powers to adopt specified private drainage assets.

Drainage of Highway

Includes collecting, transporting, and disposing of surface/subsurface water originating on or near the highway right of way or flowing in streams crossing bordering that right of way.

Drainage of highway is important because water damage highway structure in many ways. The water which are dangerous for highways are:

Rainwater: Cause erosion on surface or may seep downward and damage pavement (surface drains)

Groundwater : May rise by capillary action and damage pavement (sub-surface damage)

Water body : May cross a road (river/stream) and may damage road (cross drainage words)

Related Pages: It is more appropriate to take care of drainage at the time of location survey.Ideal location fro a drainage stand point would lie along the divides b/w large drainage areas. Then all streams flow away from the highway, and the drainage problem iss reduced to caring for the water that falls on the roadway and back slope.In contrast location paralleling large streams is far less desirable as they cross every tributary where it is largest. Ideal locations avoid steep grades and heavy cuts and fills as they create difficult.

Pavement and Shoulder

The highway engineer should ensure that the precipitation is removed from the pavement as soon as possible and that highway drainage is done efficiently. Water that falls on the road way follows laterally or obliquely from it, under the influence of cross slope. Or superelevation in pavement and shoulder.

A suitable value of cross fall for paved roads is about 3% for carriage way with a slope of 4-6% for shoulders. And increased cross fall for the carriage way e.g. 4% is desirable if the quantity of the final shape of the road surface is likely to be low for any reason,

DEVELOPMENTS IN DRAINAGE MACHINERY AND MATERIALS

Materials for subsurface drainage systems include drainpipes and their accessories, envelope materials and auxiliary structures.

PVC and PE are generally used as pipe materials for corrugated plastic lateral drains. Concrete pipes are used for larger collector drains. Specifications and standards for clay, concrete and corrugated plastic pipes are available in the USA and in Europe a standard for corrugated PVC pipes has been drafted.

Pipe Accessories

Pipe accessories, such as end caps, couplers, pipe fittings and reducers, and rigid pipes for drain bridges and lateral outlets are also available.

Design criteria and hydraulic calculation procedures with respect to pipe diameters are well known. FAO is currently preparing simple computer programs for both single drains and multiple drains, in which the pipe diameter changes as a way to reduce material costs.

Selection of Pipe Material

The selection of pipe material and size for a particular project depends mainly on local availability and cost. In this context the following questions can be considered:

(i) Are technical specifications commonly used in practice to check the quality of pipe materials used in drainage works?

(ii) The contribution of the pipe size to the total cost of the subsurface drainage system is relevant, however, in some cases no special attention is paid to drain diameter calculations and pipes are over-dimensioned, especially if long drains are designed.

DRAIN ENVELOPES

Drain envelopes restrict the entrance of soil particle into the drain, improve the hydraulic conductivity at the soil-drain interface and provide structural stability around the drain. Mineral granular envelopes and prewrapped fibrous organic envelopes have been used in the past, but currently synthetic envelopes -geotextiles and loose synthetic fibres- are being used.

Specifications and standards for envelope materials are also available, as well as rules and recommendations to predict the need of an envelope and design criteria for the selected material. However:

- methods validated by field experience to assess the need for drain envelopes;
- selection criteria for the most appropriate envelope material, depending on local soil conditions, need verification;
- case studies on the evaluation of the performance of drain envelopes in the field, especially for synthetic envelopes are important; and
- research activities required to support the above issues are necessary.

Auxiliary Drain Structures

Connection structures, inlets and outlets of water and special structures, such as cleaning facilities and structures for controlled drainage and subirrigation, are common auxiliary structures of subsurface drainage systems.

In the academic discussion, the following issues can be covered:

- Quality control and maintenance of outlet structures.
- In composite drainage systems, junction boxes and manholes are sometimes hardly used or not used at all and may be unnecessary with GPS availability. Examples with practical data about the comparison of savings in construction costs and increments in maintenance costs may be useful for future designs.
- Designs for special structures for controlled drainage in the lateral and collector outlets are available, but examples of construction and operation of such structures are not frequent.

Use of Trenchers

Trenchers of various types have been used in the past and still are used with success to install subsurface drains, especially for clay and concrete pipes and for granular mineral envelopes. Since the introduction of corrugated plastic pipes and prewrapped envelope materials, the installation speed has increased by using trenchless drainage machines.

However, although the use of the laser grade has improved the precision of installation, the control of the work quality has become more difficult, especially as far as the grade line accuracy is concerned.

Suggested topics for papers describing new developments and for discussion may be the following:

- Installation of drains under adverse conditions, such as in wet soils and in lands with shallow water tables.
- Subsurface drainage works in unstable soils.
- Checking the quality of drainage work during installation and after the drainage system has been installed.

Drainage Materials

Subsurface drainage systems adequately installed with appropriate drainage materials have low maintenance requirements. Dry rodding is sufficient to remove slight clogging, fresh ochre and roots proliferating inside the drainpipe, specially near the lateral outlet. To remove sediments and serious ochre deposits, and to clean clogged perforations jet flushing is necessary. Field experience achieved during the past years shows that medium pressure equipment is most recommended and flushing should be used only in case of dissatisfaction with or deterioration of drainage system performance.

The effective length of the flushing equipment is less than about 400 m. Therefore, this technology is effective for single drains with lengths up to 400 m and for multiple drains if manholes are constructed. However, extended laterals and composite drainage systems are sometimes installed without manholes. In these cases, are there other effective cleaning facilities or procedures?

TREES AND CROP PRODUCTION

Trees can grow on agricultural land in many useful and profitable ways. These include: windbreaks, treed fence rows, fruit and nut orchards, stream buffers, shelter for pastures, shelter for farm buildings, plantations, woodlots and reforested marginal land. Cropped farm fields and pastures are often under-drained with tile to remove excess water to enhance crop production. Some species of trees can aggressively plug farm drain tiles with roots while other tree species rarely plug drains. If conditions are favourable, it is believed that any tree species has the ability to plug farm drains with roots. Where crop production is to continue, tree planting projects on drained land must be designed to sustain the use of tile drains and not interfere with removal of excess field water.

Properly Designed Drainage System

A properly designed drainage system should quickly remove water from farm soil whenever an excess amount of water is present. Excess soil water can occur any time of year, however, is typically most abundant from late winter to early spring, late summer to early winter and sporadically during the growing season due to heavy or prolonged rains. The need for field drainage can vary from season to season. Field tiles that have remained fairly dry for several years may run water frequently or constantly during wet seasons if water tables become high. Tiles that drain natural springs will always run water. During most seasons, farmers realize production advantages by having field drains in place.

Drain lines blocked by tree roots will disrupt proper water drainage. How can this problem affect crop production? Winter melt water and spring rains may not drain adequately to allow early tilling and seeding especially on heavy soils. Wet soils are cooler and can delay crop germination and growth, reducing crop yields. Sporadic summer flooding can remain pooled too long in low spots, resulting in crop damage. In late summer and fall, soil may be too soft to permit heavy harvest equipment onto the land. It costs time and money to locate the plug and replace it with new pipe. Farmers depend on a functional drainage system to remove excess field water.

Farm Drainage

Farm drainage is a production investment that can cost $600 /acre to install. Tree plantings added to farmland can also be designed as production investments. In order to establish trees and shrubs on farmland in beneficial ways without risk of plugging drain tiles with roots, recommendations on tile drain installation are outlined in OMAFRA Publication 29, Drainage Guide for Ontario. These recommendations have been established through field experience and are recognized by drainage contractors. If questions on tile and tree root plugging arise, farmers should consult with their drainage contractor or OMAFRA Agricultural Engineer.

Precautions for stream (riparian) buffers

Where trees, shrubs and weeds associated with wet soils are established or are permitted to grow naturally along a watercourse, perforated tile that passes under the buffer to an outlet can quickly become plugged by roots. For field tiles that drain into buffered streams, intermittent watercourses or ditches, a section of non-perforated tile should be installed.

The non-perforated section of tile should extend from the outlet, pass under the vegetated buffer, and continue for at least 15 meters into the cultivated field where it can then connect to standard perforated pipe. Roots will not penetrate non-perforated pipe. Worry-free drainage of field water will continue with added benefits realized by having buffered vegetated watercourses.

How tree roots grow in the soil

Roots of trees grow to new areas of soil to increase the root surface area. Nutrient uptake, water absorption and anchorage are key functions of roots.

Roots grow proportionately in size with the above ground tree and maintain a specific root-to-shoot ratio.

Roots of many species of trees, some weeds, several shrub and crop species can grow close to and within tile drains as they expand their ability to acquire water and nutrients.

Back-filled soil over drainage tiles provides easy access for roots to the tile due to the breakup of compacted soil layers and loosening of the back-filled soil by the drainage installation equipment.

Roots do not actively search the soil for moisture and nutrients but grow more vigorously as they randomly encounter more favorable growing conditions such as, increased moisture and nutrient levels.

Root growth conditions can continue to improve until moisture becomes excessive or nutrients reach toxic levels at which point root growth declines.

The ideal amount of soil moisture or the ideal amount of each nutrient is entirely dependent upon the tree species.

Physical soil properties can become more favorable for root growth with increasing moisture. Roots may develop more vigorously towards an increasing humidity gradient and moisture gradient. Roots can push their way through soil easier towards an increasing moisture gradient as soil becomes more pliable due to increasing amounts of water.

Other soil factors such as oxygen concentrations and soil particle size also contribute to ease of root growth.

How tree roots plug drain tiles

Drainage tiles that are perforated with holes (modern plastic pipe), have gaps (sectional clay tile) or are damaged by cracks can be plugged by roots. Non-perforated pipe cannot be plugged by roots since there are no entry points.

Roots are more likely to be found within tile after a prolonged dry period as root systems expand downward to increase their ability to absorb water.

A root will likely stop growing once it enters a dry tile but can remain alive. The root will not proliferate to plug the tile if the root does not encounter a water source.

Once running water or standing water becomes available inside a tile, tree roots that are present may proliferate and plug the tile.

The rate of root growth and an ability to plug is dependent on the species of the root occupying the tile. Roots will plug tile slower if other sources of water are available outside the tile during the same period of time.

Problems of Drainage

Drainage problems are first noticed by the farmer as a wet spot in a field that does not drain as fast as it did in previous seasons or as an area of unhealthy crop. Upon inspection of tile outlets water may be observed running later in the spring and early summer. The late water flow may not be due to 'late flowing water' but can be due to a slow leak in the plug itself. A backup of water up the tile system may simply be taking a much longer time to drain.

The root growth which created the plug could have progressed the previous season in early autumn and continued as late as December. The plug may have developed over several seasons. An old tile system may be losing the ability to effectively drain the land due to accumulations of sediment or pipe collapse. Plugged sections of tile will need to be located, cut out and a new section of tile spliced into the line.

Root masses that form within tile can occasionally break free from the parent plant and travel downstream inside the tile, eventually blocking water flow at a different location. These shifting plugs have been found blocking drains of interconnected neighbouring farms causing crop damage. Determining where the root mass originated from can sometimes be difficult.

Water can run through tile constantly or it can flow for an extended duration into the growing season due to drainage of natural springs or drainage of areas in a field where the water table meets a tile. Indication of this water flow can usually be observed at the tile outlet.

Depth to the water table can vary from one season to the next and is dependent upon seasonal rainfall patterns. Tiles that are dry during average growing seasons may have late flow of water during wet seasons. Field locations having wet tile will be a risky area to establish any tree species since plugging by roots may eventually occur.

Water may be running through sections of field drainage later into the growing season but may not be realized by the farmer. Tile can drain water from an up-slope, wet area or from a spring however, as the water makes its way down the tile to drier areas the water can leave the tile through perforations. The water can re-enter the soil in another area before reaching the drain outlet. In these situations, and unknown to a farmer, trees planted close to tiles that have standing or flowing water could cause root-plugging problems.

At a low field elevation, plugging in tile by roots may encourage proliferation of roots of other trees upstream in the line since water remains present until the plug is noticed and subsequently released.

How Trees Remove Soil from Soil?

By transpiration, trees, like crops, can remove significant amounts of water from soil. If planted dense enough, could trees act as an effective natural substitute for tile drainage on farmland? Trees remove little soil water during times of critical farming activities. In order to remove soil water through transpiration, water must evaporate through photosynthesizing deciduous leaves or needles of conifers. Trees remove water from soil when crops absorb water from soil, during the growing season, from mid-spring to early autumn. At other times of the year, trees transpire very little water out of the soil. Although some moisture evaporates from trees during winter dormancy, the amount is small. Trees remove very little water from soil before early spring and after mid-autumn since leaves are absent or are no longer functional, and for most conifers, needles are either ending or beginning their winter dormant period. Early spring and fall are critical times when land needs to be drained to allow access by heavy farm equipment. Field tile will adequately drain field water at these important times.

During heavy summer rains and occasional flooding, soil oxygen levels can be depleted enough to cause many tree species to stop water uptake altogether, especially dry-site trees. Tree species that tolerate wet soil and are able to continue absorbing water during floods would not remove field water fast enough to enable nearby crops to survive the prolonged saturated conditions. Trees will not function as a natural substitute for tile drainage during sporadic saturation of farm soil during the growing season.

OPEN DRAINS

Many farmers have constructed deep, open drains on their properties. These drains are generally excavated by a back hoe or a bulldozer. Whilst the farmers are prepared to pay for and construct these drains, land-holders further downstream are usually unhappy about receiving the extra saline water that flows in the drains. They are also concerned about sediments carried in the water. The result of this conflict has been bitterness between some land-holders. To reduce these conflicts and the degradation of land downstream, drainage regulations were introduced in 1992.

During rain or irrigation, the agricultural fields become wet. The water infiltrates into the soil and is stored in its pores. When all the pores are filled with water, the soil is said to be saturated and no more water can be absorbed; when rain or irrigation continues, pools may form on the soil surface.

Some part of the water present in the saturated upper soil layers flows downward into deeper layers and is replaced by water infiltrating from the surface pools.

When there is no more water left on the soil surface, the downward flow continues for a while and air re-enters in the pores of the soil. This soil is not saturated anymore.

Our home is perhaps one of our most precious possessions. But is it enough to boast of a beautifully decorated interior? Think again! The external look of our dwelling place appeals visually but what's going on underneath the foundation is often much more important. The drainage system of any abode needs equal attention in order to provide sustainability and safety.

DRAINAGE SYSTEM

Drainage is the natural or artificial removal of surface and sub-surface water from an area. A smart drainage solution is mandatory to dissipate surplus water both in farming lands and in residential areas. Untreated excess water abets the chances of erosion of the foundation, making the structure of your home unstable. On the other hand agricultural soil needs drainage to enhance production or to manage water supplies.

An underground drainage system comprises of pipes and storage tanks made from perforated wall modules to form the desired size and configuration. They are wrapped in a water permeable geotextile. The system is preferably buried in clean sand, whereby rainwater and runoff water is directed to flow through the water permeable geotextile.

Drainage can be either natural or artificial. Many areas have some natural drainage which means that excess water flows from the farmers' fields to swamps or to lakes and rivers. Natural drainage, however, is often inadequate and artificial or man-made drainage is required.

Surface drainage is the process of removal of excess water from the surface of the land. This is normally accomplished by shallow ditches which are also called open drains. The shallow ditches discharge into larger and deeper collector drains. In order to facilitate the flow of excess water towards the drains, the field is given an artificial slope by means of land grading.

Subsurface drainage is the removal of water from the root zone. It is accomplished by deep open drains or buried pipe drains. Excess water from the root zone flows into the deep open drains. The disadvantage of this type of subsurface drainage is that it makes the use of machinery difficult. Pipe drains are buried pipes with openings through which the soil water can enter. The pipes convey the water to a collector drain.

Drainage system that lies beneath your home works hard at all time to keep your abode prim and proper. Not only does it have to process water waste that is created by bathing and washing dishes, but it also handles sewage waste and outside water that is draining off your home. This is a tough set of responsibilities for the residential underground drainage system to handle so you might want to think proactively about keeping this system in working order.

Under these regulations "farmers must notify the Commissioner of Soil and Land Conservation of their intention to drain or pump water for the control of salinity" so that anyone likely to be affected is informed.

In April 1995, Agriculture WA undertook a stock-quality drilling program in the North Stirling area. Three of the selected sites were on Mr. Ken Pech's farms which form the headwaters of the Six Mile Creek. Ken showed us the deep drains he had constructed more than ten years earlier. His photos showed that there was an apparent reduction in salinity in an area between 100 and 200 m away from the drains. The improvements were unusual because similarly constructed drains in other landforms have had little or no effect on the salinity status of their surrounding areas. We decided to evaluate these drains and present the results for the benefit of other land-holders that may want to construct deep drains.

Methods

Geological description of the area were based on:

(a) soil profile information from three bores which were drilled. the drilling logs for two of these bores, which are typical of the profiles in the area.

(b) the rocks found in the area were Pallinup Siltstone and spongolite, both of which are of Tertiary origin, and a few Precambrian basement rocks (found in the upper section of the drain).

(c) descriptions of soil profiles along the drain

We recorded the position of major seepage zones which were along the mid section of the drain.

Aerial photographs from 1985 and 1993 were interpreted to find the extent of saltaffected land before, and eight years after, construction of the drains.

The length of the drain was measured from the 1993 aerial photographs. We contacted a local contractor to obtain current construction costs. These quotes are related to the size of drains.

In December 1996, we walked along the drains and:

- Noted which areas were strongly salt-affected (had become bare grounds);
- Used a Geonic EM38 instrument to define the salinity status of the suspected areas (using the relation given in Ferdowsian and Greenham, 1992);
- Described four typical soil profiles ,
- Estimated the flow rate at several sites ;
- Sampled the baseflow to measure its salinity.

Results

Geology and Soil Material

The area is covered by Pallinup Siltstone which is a Tertiary Formation. The Pallinup sediments in this area, are mainly composed of fine silt, clay and occasionally fine to medium sand. The siltstone, in some areas, contains pink and white spongolite (a siltstone full of sponge spicules).

The Pallinup Siltstone overlays the basement rock which is of Precambrian origin and part of the Yilgarn Craton. There are basement (granite and gneiss) highs in the area that are covered by sediments. Weathered granite can be seen along the bed of the main drain.

Drain Specification

The drainage system is composed of one main drain (6.5 km) and 7 km of lateral ones. The main drain and a few of the laterals (8 km altogether) were constructed in 1984. Another 4.5 km of lateral drains were constructed in 1985 and 1986. Only one lateral drain (about 1 km) was constructed after 1993 and so this cannot be seen on the 1993 aerial photographs.

All drains have a trapezoidal cross-section. The bed of the main drain is 3 m wide and its sideslopes 1:1 (vertical : horizontal). The beds of lateral drains are 1.2 m wide and their sideslopes 2:1. With the exception of approximately 100 m; the granite high near M6), the drains are entirely in the Pallinup Formation.

Condition of the Drain

With the exception of a few places, the sideslopes of the lateral drains are stable have not collapsed. The large piles of spoils are bare and exposed to erosion.

Coarse grains of sand could be seen along the banks of the lateral drains which indicate that the spoils from the lateral drains have been eroded and entered the drains.

Some sections of the main drain were badly eroded. Gullies, 1 m apart and up to 0.5 m deep, had formed along the sideslopes in some sections of the main drain. Part of the eroded material had filled the base of the main drain and part was transported out of the area. There were also indications of sedimentation further downstream. A typical site where this silting may be seen is 500 m before the creek crosses the South Formby Road. On the 1985, aerial photographs there are four pools in this area. These pools are absent on the 1993 aerial photographs. It is likely that these pools have been filled with silt between 1985 and 1993. Silting of the creek bed can also be seen further downstream, east of the South Formby Road.

Flow rates, baseflow salinity, evaporation along the drainage system and salt export. The bed of the drain was wet throughout its entire length but only a few seepage areas which had sandy profiles contributed to the baseflow. Seepage zones included sections of the drain which had been cut into the floor of salt lakes.

The flow rate along the main drain (December 1996) varied between 0.2 L/sec and 2 L/sec (M1; at the outlet of the property). A few of the lateral drains had no or little flow. The flowing water was baseflow which was ground water seeping into the drain.

Baseflow salinity varied between 5600 and 8400 mS/m. The lowest salinity reading (5600 mS/m) was at site L5 which is close to a seepage zone.

This salinity is probably similar to the salinity of the groundwater. The salinity of the baseflow at the outlet of the property was 8300 mS/m. The increase in salinity is probably due to evaporation.

The ratio of baseflow salinity at the outlet of the property (8300 mS/m) and the lowest salinity (5600 mS/m) shows salt concentration due to evaporation along the drainage system. Thus without evaporation, the flow rate at the outlet of the property (Q) would be:

$$Q = 2 \text{ L/sec} \times 8300/5600 \text{ or } Q = 3 \text{ L/sec}$$

The base flow was carrying approximately 7.9 tonnes of salt out of the property each day in December 1996. We expect that the salt export will probably be much higher during winter months. Silberstein (1989) found that the average salt export, at the outlet of the drainage system, was approximately 240 tonnes per day, between

10 th May and 30th July 1988. The annual salt export figure was 120 tonnes per day during 1988. He attributed the large salt export to the excessive runoff which was the result of an exceptionally wet season. The total rain falling during May, June, and July 1988 was 287 mm, compared to the 10 year average of 144 mm for the same period.

Lewis (1992) studied the North Stirling Basin, which is west of the study area. Average salt storages in the 13 bores which she studied were 1830 t/ha (average depth 13.50 m). Salt stored in the drain's catchment (9600 ha) would probably be the same as reported by Lewis.

Assuming that the drain would continue to remove 7.9 tonnes of salt each day, it will take more than 6000 years for the drain to remove all of the salt from the soil profile. Even with the highest salt export rates (120 tonnes per day; in 1988), it would take about 200 years to flush half of the stored salt from the catchment of the study area.

Water balance and potential salinity

We expect that between 10 mm (in a year with average rainfall) and 50 mm (in a decile 8 year; two wettest years out of ten) of the annual rainfall would recharge the aquifer each year. If we assume that the catchment boundaries are the same as the aquifer boundaries, then the total area that contributes to the base flow in the drain will be 9600ha. The annual base flow out of the property and evaporation along the drainage system is 94,600 m 3 (3 L/sec) which is less than 1 mm per year over the total catchment area. The rate of base flow discharge is very little compared with the recharge under pastures and crops. Thus, the deep drain alone will not stop soil salinity and many more areas may eventually become salt-affected. As an example, area 1, has become saline since 1993.

Effect of the deep drain on the extent of the salt-affected area

In this report, salt-affected areas are defined as lands that have become bare and appear as highly reflective areas on the aerial photographs. These areas do not include moderately or slightly salt-affected areas (barley grass areas or salt-affected cropping land).

About 251 ha of land was strongly salt-affected on the 1985 aerial photographs. These areas looked continuous and spread out. Most of the affected areas were in the natural drainage beds.

The total salt-affected area based on 1993 aerial photographs was about 224 ha. By comparing the 1985 and 1993 aerial photographs, we could see that:

- Salt-affected areas close to the deep drains had been reduced, and in some places had become isolated ;
- During the same period, salinity had encroached into other areas that were further away from the drain.

We found that the salt-affected land in the study area had reduced by 27 ha between 1985 and 1993. Out of the 27 ha which showed signs of improvement, about 8 ha was cropped in 1996 and the balance (19 ha) was within the natural creek beds and flood plains.

Improvement of creek beds and flood plains could be partly attributed to them being fenced off and revegetated with salt-tolerant species. These areas had been mounded and planted with salt-tolerant species. Melaleuca thyoides, Acacia saligna and Atriplex spp. were among the surviving plants. Melaleuca thyoides was very impressive with a large canopy (5 m in diameters) and healthy growth.

The improved area, furthest from the drain which was cropped in 1996 was a 3 ha patch of land) up to 200 m away from the main drain. Improvement of this land is entirely due to the effects of the drain since there is no treatment other than the drain. The area was bare in 1985. This area had been cropped but not harvested because the crop (barley) looked patchy.

The Geonic EM38 readings on this area were between 80 and 120mS/m. These reading showed that the area was still moderately salt affected. This land is probably on a sandy lens that crosses the deep drain. By having a high hydraulic conductivity, this sandy lens has extended the effect of the drain far beyond the expected distance. We observed one section of the deep drain that had higher seepage rates than other areas and found that the sandy lens crossed the drain here.

Feasibility of the Deep Drain

The Net Present Value (NPV) is a measure used in economic analysis to determine a project's net benefits. It is the discounted value of expected benefits, less the discounted value of expected costs. The Net Present Value of the drain has been calculated based on the following assumptions:

- Drain construction cost $30,000 (1985);
- No maintenance until year 2000;
- Benefits from construction of drains commence in 1989;
- Barley grown on reclaimed area;
- Barley gross margin $130/ha.

The analysis indicated that at a discount rate of 8% the investment cost would be recovered if 40 ha of land was reclaimed. The actual reclaimed area (8 ha) will not recover costs over the life of the project. The area required to break even varies with the level of discount rate.

The analysis does not account for off-site effects of the drain such as export of salt and silt to the other properties. It also does not account for intangible gains or losses such as the scenic values of reclaiming land, look of the drain with its spoil mounds and access to paddocks.

Unhygienic and Unsafe Open Drains

A huge open drain. manholes that seem to be sinking, which keep overflowing at intervals. Roads that get flooded with sewage when it rains. A perennial stink. This can best describe Palmgrove Road.

It's the broad road off the busy Victoria Road in the Central Business District, where the real estate rates are steadily climbing. And yet Palm Grove Road has been neglected by the civic authorities. The old area had its sewage lines laid over 100 years ago.

The ancient, leaky pipes are not just corroded but clogged with garbage and sewage does not flow freely through the lines. The road is dotted with manholes, which overflow every now and then. Many manholes on the perpendicular Victoria Road have begun to sink. It gets worse when it rains. The drainage system fills beyond its capacity and sewage mixed with rainwater floods the road and many apartment complexes that line this road.

Residents of the road have covered many manholes with large stone slabs to keep the lids in place. This, however, does not stop the sewage from flowing out on to the road. There is a huge open drain along the road. It crosses the breath of the road in the middle and flows along the other side of the road.

The drain is left open. The water level in the open drain is high and heavy rain and an assortment of garbage is enough to fill it up. Residents living in the apartments facing the drain requested the BBMP several times to close the drain. They were told that the drain could not be closed, as de-silting work was planned. "We did not see any de-silting happening, but the drain remained open. It is unhygienic and unsafe. Though there is a wall and a fence along most parts of the drain, the wall is not high and children might fall into it. Moreover, it is an ugly sight.

However, saturation may have lasted too long for the plants' health. Plant roots require air as well as water and most plants cannot withstand saturated soil for long periods However, rice is an exception.

Besides damage to the crop, a very wet soil makes the use of machinery difficult, if not impossible.

The water flowing from the saturated soil downward to deeper layers, feeds the groundwater reservoir. As a result, the groundwater level.

DRAINAGE OF BARNYARDS & CATTLE-LANES

The stockman is familiar with the difficulties to be met in keeping ground which becomes puddled by the frequent tramping of live stock from becoming exces sively muddy. Such ground cannot be materially benefited by placing tile drains underneath the puddled surface. The remedy consists in preventing all water from outside sources from finding its way to the yard, leaving only the direct rainfall to be contended with.

The roof water from all buildings adjoining the yards should be taken care of by eaves-troughs, and downspouts which should conduct the water either into cisterns or into a tile drain provided for the purpose.

This receiving drain should be laid around the buildings and discharge into some open channel or into a system of field drains. A tile of 8-inch diameter will carry the roof water from a large barn, and may discharge into a field main without overcharging it, for the reason that in all heavy rainfalls the roof water will have passed through the drain before soil water will have had time to enter. It is not desirable that stock-yards should be kept dry by surface drainage unless the value of the manure is to be disregarded. It is the practice of many good farmers to so arrange the stock-yards that all rainfall will gravitate toward the center, which thereby becomes the receptacle for valuable manures and gives drainage to the outer parts of the yard. In all cases the surface water from surrounding land should be cut off by shallow trenches supplemented by under drains.

These suggestions, if followed, will result in a great amelioration of the mud evil so often endured by farmers and stockmen under the impression that there is no remedy for the knee-deep conditions of their yards.

No surface water should be permitted to enter tile drains direct unless precautions are taken to prevent mud and debris from entering the drain. It is frequently desirable to remove surface water by means of underdrains from certain places which are not susceptible to drainage by soil filtration nor provided with surface-drain outlets. Some of these are ponds or depressions by the roadside, yards which are kept closely compacted by constant use, drainage from carriage washes and barns which is charged with muddy material, and other like necessities.

For such purposes, the catch-basin will serve an excellent purpose. It is a well constructed with brick, 3^ feet in diameter, with a depth of 2 feet below the outlet pipe for the settling of mud and heavy material which should be removed whenever the space below the discharge is filled. The discharge pipe should not be less than 6 inches in diameter in any case and should not be connected with any extensive field system, but should extend direct to some large outlet. The inlet should be 10 or 12 inches in diameter and have two rods placed vertically through drilled holes near the entrance to serve as a screen. Both inlet and outlet pipes should be vitrified sewer-pipes to withstand the effects of freezing, which common clay ware will not. A wooden box instead of a brick well will serve the same purpose, but must be renewed when the wood decays, and is' obviously less permanent as an improvement. In some respects, however, the wood construction is more desirable.

The inlet may then be a hole in the side of the box at the surface flow line, with iron rods fastened vertically across it.

A serviceable and easily constructed catch-basin may be made of the sections of 24-inch sewer-pipe, which should be set end to end in a vertical position and the socket joints secured by cement mortar. The bottom section should be a straight pipe, the upper two sections should have T's, one of which should be used for an outlet and the other for an inlet. A 2-inch plank cover fitted and dropped into the top socket finnishes a neat and desirable catch-basin.

Catch-basin constructed of Sewer-pipe. These catch-basins will require some personal attention at times. Straws and other surface rubbish from the surface ditch will gather against the inlet and must be removed. The mud which accumulates in the bottom of the basin should be removed from time to time. It is, however, a useful and convenient accessory to drainage work.

PROCEDURE TO CONSTRUCT TILE DRAIN

Cellars which are excavated in clay or loam soils usually become wet and require drainage. A common method of procedure is to construct a tile drain to carry the water away after it enters the cellar. Another is to cover the walls and bottom with a thick coat of cement mortar to prevent water from entering. If ihere is much soil water to contend with, it frequently bursts through the coating used in the latter method, and when the former is used, the cellar remains damp even when the drain removes the free water which percolates through the soil and finds its way to the low point in the bottom from which drainage is made.

The proper plan to follow is to prevent water from entering by means of a tile drain, which should be laid entirely around the building 4 or 5 feet distant from the walls and nearly if not quite as deep as the floor of the cellar. This drain should be of 4-inch tile laid on an even fall of one inch in 16 feet, and connect with a main which will carry the drainage safely aay from the house. By this method the ground about the house as well as the cellar will be kept dry and wholesome, and is the best known plan for securing a dry cellar for a country house where natural drainage is deficient.

The underdrainage of the ground occupied by buildings and surrounding yards and gardens is often neglected on the supposition that these grounds have sufficient natural drainage, which is not always the case.

Filtration & Water Supply

The filtration of all surface water through the soil of the lawn, garden, and surrounding grounds in general, and its removal by the process of underdrainage, will add much to the ease with which walks may be constructed and maintained, and to the satisfactory growth of all useful and ornamental plants which contribute so largely to the beauty, value and healthfulness of a country residence.

The water supply is usually taken from a well located near the residence. While the supply of this well may have its source in a vein of clay, and be all that is desired in point of purity and coolness, it is subject to contamination by surface and soil water which percolates through the earth and finds lodgment in the well. Some deep under drains laid about the well will prevent the pollution of the water from this source where the wells are located on level or undulating tracts of land.

BARNYARD DRAINAGE

Nothing can be more discouraging to both beginning and experienced graziers than having to cope with muddy cows. But the good news is there are many cost-effective ways to keep cows clean, even during soggy spring weather.

While mud is a particular concern for dairy farmers, beef and sheep producers also need to pay attention to proper laneway planning and management says Rob

DeClue, conservationist with the Chenango County SWCD and Graze-NY technical specialist. "You can have great forage ready to graze. But if you can't get the animals to the paddocks, it won't do you any good," observes DeClue. "For many graziers, laneways are the weak link in their grazing system, and when the going gets muddy some of them give up."

Many of the potential benefits of grazing — such as less prep time before milking and fewer foot- and udder problems — won't be realized if your cows have to waste energy slogging through the mud getting to and from the barn, he adds. "Plus the general public — which is concerned about animal welfare — associates clean animals with happy animals," he observes.

DeClue offers these farmer-tested tips:

Start at the barn door: What good are great laneways if your barnyard is a mess? Be sure to channel roof water and divert other runoff away from loafing areas.

Invest up close first: Focus your efforts on high-traffic laneways close to the barnyard. Management — not money — might be a better way to tackle wet spots farther out in your grazing system. For example, plan to mechanically harvest paddocks that may be tough to reach in spring without forcing your stock through muddy spots.

Plan ahead: When setting up your grazing system, skirt wet spots when you locate your laneways. Recall places where existing pastures were slow to drain or cropland was often too wet to work, and route laneways around them. Where your stock must cross draws, intermittent streams or hollows that are wet after storms, consider building a stabilized, durable crossing.

Turn the problem into the solution: Where hardpan soils limit drainage, you can often remove the topsoil, bulldoze the subsoil to form a raised lane, then grade and crown it so that surface water flows away from the path.

Beef-up high traffic areas: Lanes close to the barn take a constant beating. They may require more intensive work, says DeClue. He recommends this strategy: First, excavate the top foot or so of soil from the lane. Then line the area with geotextile fabric, fill with coarse aggregate (such as clean bank-run gravel or equivalent), crown above the original grade, top with 1-inch or so of mixed-size crusher dust or lime dust, then roll. The dust acts as a weak cement to help bind the larger aggregates. The cohesive surface helps prevent cows from tracking stones onto concrete pads where they can cause stone bruises.

In a recent posting on the graze-l internet discussion group, Ohio dairyman F.W. Owen described how he avoids stone bruises by substituting the tabs punched out from asphalt shingles, available free for the hauling from a local factory. Other farmers suggested cushiony "chips" made from old tires that are often available locally. But they also cautioned to avoid those that may be contaminated with metal fibers from steel-belted tires and to check with local environmental authorities first.

Many farmers are tempted to skip the geotextile fabric, which keeps the animals from punching the aggregate down into the soil. But in DeClue's experience, the relatively low cost of the material (usually about $.60 to $.90 per running foot for 12.5-foot wide stock, depending on the type used) is worth it, in most cases. "Barring native gravel, a tough hard pan or bedrock near the surface, it's just a matter of time before the aggregate will get swallowed up, and you'll have to haul in more," he explains. As the muck oozes up from below, animals also are more prone to bruise their feet on the embedded stones.

Rotate laneways: Many farmers have minimized mud problems by building double-wide laneways close to the barn, splitting them down the middle, and alternating from side to side as needed. Sometimes it might be necessary to form a temporary lane with portable fence along the inside edge of a paddock or major subdivision to route cows around a wet spot in the permanent laneway.

In the far reaches of your grazing system, DeClue suggests building a single permanent "backbone" fence instead of two laneway fences. Form a lane with temporary fence on one side of the backbone to reach paddocks on that side, then flip the lane to the other side to reach the other paddocks. On remote laneways that don't get much use, the sod may be tough enough to support cows even when it's wet, assuming that they'll only need to use the lane a few times before extended rest.

Watch your water: Locating water tanks in your laneway increases traffic in the lane and can lead to more mud problems. Water distribution systems that feed movable tanks connected to strategically placed hydrants with 50 to 75 feet of above-ground hose allow you rotate and rest areas muddied up by the animals.

Watch your weight: Keep tractors and heavy equipment off of your livestock lanes — especially when it's wet — or build a heavier-duty lane designed to take more punishment.

Seek help: Technical assistance and cost-sharing is available for many of these strategies.

"Sometimes the money you spend making improvements doesn't look good going out," says Lee Wilson, dairyman and NYPA chair. "But the cows sure look better coming in."

Kill two birds with one stone peter Mapstone, who grazes 100 cows and heifers in Manlius, N.Y., found a creative solution to mud problems that keeps his cows' feet dry and their thirst quenched. Mapstone had a neighbor skim the clay off a perennially wet spot at the top of a hill, excavate the underlying shale to form a nearly half-acre pond, and relined it with the clay. After the 12-foot-deep pond filled, Mapstone used 1.5-inch plastic pipe to gravity feed the water to his pastures and barn. The neighbor also formed a half-mile 1- to 2-foot thick lane by grading and rolling the excavated shale. "The shale is sharp but it's pretty soft. There's a lot of dirt mixed in with it so it's a lot like crusher-run. The roller packed it down hard with no sharp edges sticking up," says Mapstone. "It works great. The cows never sink in more than an inch." Mapstone also used some of the shale in his barnyard, and sold some to help pay for the project.

Managing Outdoor Confinement Areas and Livestock Yards

An outdoor confinement area (OCA) is defined by the regulation as an enclosure for livestock or game animals that has all of the following characteristics:

- an unroofed area, with the exception of small wind or shade shelters that are under 20 m2(200 ft2). Such shelters are considered part of the OCA
- permanent or portable feeding or watering equipment
- a grazing or foraging area that accounts for less than 50 per cent of the animal's dry matter intake
- fences, pens, corrals or similar structures to confine the animals that are either permanent or temporary
- access to a barn. The non-roofed area may be an OCA.

Some livestock yards or paddocks might not meet all of the characteristics of an OCA. For example, grazing systems, including crop residues, and true pastures are not OCAs.

Permanent OCAs

A permanent OCA (POCA) is a site with the characteristics listed above where the animals:

- have access to it for at least 200 days/year or 2,400 hours in a 365 day period; or
- are present less than 200 days, but there are 300 or more nutrient units (NUs) annually in the area and stocking density is greater than five NUs per hectare on an annualized basis (5 NU/hectare/yr.) in the confinement area.

The Regulation defines nutrient units and lists (in Table 1) various species, animal sizes and the number of each it takes to generate one NU. See the nutrient management section of the ministry website for more information as well as the complete nutrient management tables for the Regulation. Below is an example of nutrient unit calculations.

Table 1. Sample Calculations of Nurtrient Units

Livestock	Nutrient unit conversion factor (from Table 1 in the NM Tables)	Nutrient Units
100 beef backgrounders	3	33
10 medium frame horses	1	10
Total Nutrient Units	43	

A benefit to fitting under the definition of a POCA is that the rules related to siting, management of manure and snow containing manure are governed by Part VII of the Regulation.

The sheep have access to the adjacent barn whereas the cattle will rest on the mounded manure and straw mound in the background. While there are other notable differences, such as a concrete floor for the sheep compared to the cattle that have a dirt floor, both pictures represent permanent OCAs.

Livestock Yards that are Not Permanent OCAs

There are many livestock yards that do not meet the legislative criteria of a POCA. Examples include:

- paved or concrete exercise yards for dairy cows where no feeders or waterers are present
- winter feeding areas with a soil base for beef cows where the cows are present for less than 200 days per year
- corrals and livestock handling areas where livestock are not fed or watered.

In these situations, the following rules apply:

- if the Regulation requires the farm to have a Nutrient Management Strategy (NMS), all nutrients must be managed according to the NMS.
- the handling of runoff from the paved livestock yard and POCA must follow Regulation, Sec. 81.
- a permanently vegetated area can only be used for runoff management on a paved livestock yard if the manure from the livestock has a dry matter of 30 per cent or greater as listed in the Nutrient Management Tables in the Regulation.

While the Regulation stipulates that manure and runoff from the winter-feeding areas listed above, as well as corrals and livestock handling areas must be managed, it does not specify how this must be done. It is the NMS that will outline how nutrients from these yards are to be managed in order to minimize any negative environmental effects.

The NMS should detail best management practices, proper siting recommendations (i.e. distance to surface water and wells) and other appropriate measures for the particular livestock yard.

Farm operators must be prepared to explain this to the Ministry of Agriculture, Food and Rural Affairs during approvals, or to Ministry of the Environment officials during inspection.

A Solid Base for Groundwater Protection

The Regulation does not require outdoor confinement areas or livestock yards to have concrete or paved floors. However, it is recommended. This allows:

- Easier collection of manure to later spread on fields that require nutrients
- Runoff to be directed to a proper area for management.

It is much easier on the operator, equipment and animals if a concrete or hard base is present in these areas, especially during wet periods.

Sensitive Features

New–or expanding–permanent outdoor confinement areas on farms with a NMS must be sited to ensure proper setbacks from sensitive features, including wells and field drainage tiles. Setback distances reduce the likelihood of manure or nutrients reaching watercourses and wells. The distance must be measured from the closest point of animal access in the POCA to the setback feature.

Table 2. Setback Distances Required for POCAs

Water Source or Feature	Setback Distance Requirement
Drilled wells of 15 m (50 ft.) or more depth and water tight casing to a depth of 6 m (20 ft.) or more	15 m (50 ft.)
Other wells including oil, gas, and water, unused and test wells not meeting the above criteria.	30 m (100 ft.)
Municipal wells	100 m (330 ft.)
Field drainage tiles	15 m (50 ft.)

If the runoff management includes using a permanently vegetated area, the flow path distance should be at least 150 m to surface water or tile inlet. The Regulation prohibits direct livestock access to surface water in some POCAs.

Runoff Control Requirements

Runoff from a livestock yard usually contains some manure. Runoff yields can range from 15 per cent to 50 per cent of precipitation and depend on storm intensity, pen slope, manure pack condition, stocking density as well as other factors.

The Regulation (sec. 81) lists the five acceptable runoff management systems allowed for permanent outdoor collection areas: roof and up-slope water diversion; runoff containment; vegetated filter strip systems; sewage works under the Building Code or Ontario Water Resources Act and permanently vegetated areas. These may be used separately or as a system.

1. Roof and Up-Slope Water Diversion—Reducing the amount of clean water entering the POCA is an acceptable runoff management method (Part VII, Regulation), along with roofing or permanently covering the area. One technique is to direct clean water from eaves troughs and gutters into non-perforated tile, berms or ditches.

2. Runoff Containment—A new or expanded runoff collection and storage system must meet the Siting and Construction Standards (Part VIII, Regulation) for liquid storage facilities. The choice of design (i.e. earthen lagoon or concrete tank) will determine the level of site investigation required.

Manure and Snow Removal and Storage

The Regulation, s. 60, allows a pen floor to be used to store manure. As a result, the NMS can allow manure to either be stored in the POCA for up to one year or placed in storage (temporary field storage or another permanent storage). It is possible to have the manure remain in the POCA for more than one year if it is declared in the NMS as part of a mounding system for animal management (resting areas). It may also be directly applied to land, or transferred to another farm unit, another farmer or a manure broker as with other manures and storages.

Snow and ice pack are a concern in livestock yards as the material is bulky and when it melts can lead to increased runoff volumes. Options for managing this 'feedlot snow' material include:

- Transferring to a manure storage facility.
- Transferring to a temporary field storage site, if the NMS permits a method for dealing with melt water runoff.
- Land application, if the material meets the parameters in Part 11 of the NM Protocol. and
- There is a 40 m setback from the top of the bank.
- A minimum application setback of 60 m to a drilled well, 400 m to municipal well and 120 m to all other wells.
- The maximum slope of the field is three per cent.
- There is minimum 6 m vegetated buffer along all surface water in the field and.
- The manure is applied at an application rate of half of the normal maximum rate (i.e. if the normal rate of application is 36 tonnes/hectare (16 tons/acre), then the rate for snow containing manure is 18 tonnes/hectare (8 tons/acre).

The Regulation does not specify requirements for applying snow-covered manure from livestock and wintering yards that do not meet the criteria of a POCA. Nor does it indicate requirements for mounding manure.

Producers with livestock yards and winter sites need to practice good management to avoid an adverse effect from the manure and runoff from these areas as well.

ROAD DRAINAGE

With a rural road network drainage of the road pavement will mostly be just down to the maintenance of ditches adjacent to the highway, and ensuring that all culverts are free flowing, not complicated but it is surprising how often these simple maintenance procedures are neglected.

One must keep the water-table low below and surrounding the road pavement to prevent the moisture content of the subgrace increasing, and hence decreasing the subgrade strength.

A highway gully is a drainage pit covered by an open metal grating located on the road edge. It's purpose is to drain rain water from the three components of adequate road or track drainage are surface, side and cross drainage.

SURFACE DRAINAGE

Road crowning

Crowning provides a low-grade fall enabling drainage from both sides of the centre of the road. This method is only effective if the crown is slightly higher than the natural surface.

Road Crowning Should be Avoided in Areas

Those places where water naturally crosses the road such as broad drainage floors. Floodways are required in these cases.

Infall and Outfall Drainage

When roads are built across the slope consideration must be given to taking water from the up slope side of the road to the down slope side of the road. When you install cross drainage you must make sure that it does not cause erosion of the road surface.

Crossfall/Outfall Drainage

The simplest method is by providing the road surface with a crossfall in the same direction as the slope (outfall drainage), thereby directing water over the road surface to disposal areas on the lower side of the road.

The other method is by providing the road surface with infall drainage back into the slope, directing water back to the up slope side of the road. If infall drainage is necessary then table drains, culverts or inverts need to be constructed. These will safely direct outfall drainage is preferred to infall drainage as there is generally no need for other drainage works such as culverts, inverts, table and mitre drains.

When installing outfall drainage on steeper slopes, batters on the downslope side of the road must not be too steep. Steep batters may erode, impacting on the road itself.

The crossfall of the road surface should be kept as flat as possible to ensure good drainage. For outfall drainage it is recommended that the maximum crossfall slope be in the order of 1.5 to 2%, whereas in fall drainage slopes can be as great as 4%.water to the down slope side of the road.

SIDE DRAINAGE

Table Drains

Table drains are excavated open channels that are built parallel to roads and tracks. These drains direct runoff to disposal areas further downslope. Table drains should only be used when natural run-off is not possible.

Fill obtained from constructing table drains can be used to build up road surfaces. The design of table drains depends on a number of factors, including the size and nature of the catchment, the slope and water volumes and flow. Larger table drains may need to be designed by engineers or soil conservation officers. Table drains should be constructed with a flat bottom (trapezoid shape). In general they should be 0.5 to 1.0m wide at the base. Avoid using V-shaped drains as they may cause erosion in the channel.

Where possible table drains should be revegetated as soon as possible after construction, and regularly slashed.

Mitre Drains

Water should be taken out of table drains at regular intervals using mitre (offshoot) drains. Mitre drains take runoff out of table drains or directly off road shoulders where table drains are absent. These drains dispose of water in areas away from the road.

Mitre drains stop water accumulating in table drains or on the road shoulder. Ideally mitre drains should be constructed so that they have a broad flat base at least 1m wide. Mitre drains also should not be graded to produce a V. Mitre drains should slope to direct the flow of water away from the road. To minimise erosion the slope should be no greater than 0.5% on erodible soils or 1% on stable soils. Mitre drain outlets effectively concentrate runoff, for this reason they should be located in stable undisturbed areas.

Mitre drain spacing is dependent on:

- the grade of the table drain or road;
- soil type and erodibility; and
- rainfall.

CROSS DRAINAGE

Engineered, stable cross drainage such as inverts, floodways or culverts can be used to collect water from upslope table drains, or drainage lines. It is generally more economical and practical to ford drainage lines using floodways or inverts than to use major culverts or bridges. On steeper country, where creeks and drainage lines are deeper, culverts may be more practical.

Inverts and Floodways

Care must be taken in the design and construction of flood ways and inverts in order to cause minimal interference to natural flows.

Inverts and flood ways are designed to be temporarily over topped by water flow and minimise bank and bed erosion. They should be sited at low points in the bank and at right angles to the direction of flow.

Inverts

Inverts should be constructed with the finished surface at, or just below the level of the existing stream bed. Construction of an invert is generally based on excavating soft, erodible material. At least 300mm should be removed, geotextile may be necessary as a base.

Excavated material is then replaced with compacted granular material to provide a trafficable surface.

Floodways

Floodways are usually elevated above the bed level of the channel and often incorporate culverts to take "normal" flows with the road only being overtopped during flood events.

The design should have ends of the structure that are well anchored into the banks and obstruction to flow kept to a minimum by using gentle batter slopes on the up- and downstream faces. When it is necessary to construct an elevated floodway it is recommended that specialist advice be sought. As floodways are generally elevated above bed level protection works are required on the downstream side of the floodway to prevent erosion.

CULVERTS

When culverts are used they should be angled downward at between 1 and 3%. This will minimise silting of the pipe and prevent excessive scouring at the outflow. On drainage lines the culvert should be keyed into the streambed by digging a trench and seating the culvert into it.

The area below the outlet will need protection to prevent erosion. This protection can be achieved by armouring (eg: rock mattress) the drain downstream of the outlet, or by constructing a dissipating device.

Protection may also be required at the inlet. The location, spacing, size and type of culvert may vary. Advice should be sought from Soil Conservation Officers prior to construction.

WHOA BOYS ON VEHICLE TRACKS

Whoa boys can vary in size. They can be a couple of metres long and only 10–30cm high on walking tracks, or they may be large, gently sloping banks up to 30-40m and up to 3 m high on deeply eroded areas.

Whoa boys can be constructed in two ways:

1. By cut and fill –Lines are ripped across the area at a grade of 0.3 %. A shallow channel should be cut along this line.

Excavated material is dumped on the down slope side of the channel, then compacted and smoothed out to form a bank with even batters and a level top.

2. Using imported soil material to construct a bank with a grade of between 0.3 and 0.5% along the up slope edge of the bank. To aid trafficability, an approach and departure ramp can be cut into the bank.

The bank should be run off into undisturbed vegetation or into an existing drain (care needs to taken to ensure that erosion does not occur where the water runs down into the drain).

Rainwater, which strains through the permeable surface and is absorbed by the structure supporting the road, will damage the stability of the road itself. In these cases, road drainage must be laid longitudinally along the roadway edge and in the centre of the traffic divider to remove water infiltrations.

It need operating so that the underground water is collected and eliminated by the drainage pipe, the only effective drainage road to guarantee permanent results.

Failure Due to Hydraulic Pressure

Once water has entered a road pavement, water damage is initially caused by hydraulic pressure, i.e. vehicles passing over the road pavement impart considerable sudden pressure on the water present in the road pavement, this pressure forces the water further into the road fabric and breaks it up, this process can be very rapid once it begins.

Water that has entered the road pavement and is subject to the process of freezing (expansion) and thawing during the winter also brings about the swift failure of the road pavement.

Eventually the water will descend to the subgrade layer below the road pavement and weaken this layer thus lowering the CBR of the subgrade which the road pavement design was based upon, and deep seated failure of the road will begin.

Failure Due to Binder Stripping

Most aggregates have a greater affinity for water than they do bitumen, and with the presence of water and movement of the aggregate it is quite possible for the binder film on the aggregate particle to be broken and water to come in to contact with the aggregate surface.

Once the integrity of the binder layer has been broken it will depend upon the chemical nature of the aggregate particle and the viscosity of the binder as to how long it will be before stripping of the aggregate particles becomes an engineering problem.

Depending on the viscosity of the binder and the thickness of the binder film surrounding the aggregate the stripping of the bitumen will occur hardly at all, fairly slowly or quite quite quickly.

This can be compared :

(a) to a film of highly viscous 50pen. grade bitumen enhanced with cellulose or mineral fibres which will create a thick stable coating of bitumen around the aggregate particle, to

(b) a low viscosity, highly workable, low temperature, hand-lay cutback material for emergency patching or very low stress sites.

The second example strips quite readily, and in most cases should only be regarded as a temporary measure before being removed and a more stable bituminous material used to replace it.

There are many degrees of stripping between the two examples I quote, and it is a fact that some aggregate sources are more prone to stripping than others.

The first example I mention with the more viscous (stiffer) bitumens will display differing levels of stripping due to water action due to the nature of the bituminous mixture.

The more durable bituminous mixtures, i.e. hot rolled asphalt and stone mastic asphalt are designed and produced to have high bitumen contents with almost all voids filled with bitumen to provide an impervious matrix, that will be highly resistant to stripping.

With proprietary thin surface course systems (Thin Surfacings) surface courses (wearing courses), many of these surfacing materials will actually be designed to allow the passage of water through the material matrix.

And, even though they will contain penetration grade bitumens, because of their exposed position in the road pavement structure these porous bituminous mixtures will have increased exposure to water in the internal matrix of the material, and this will increase the potential rate of stripping of the bitumen.

The will be dependent upon thickness of binder film on the aggregate particles, this usually being closely related to the binder content of the bituminous mixture.

This is when the pavement has failed, maybe not completely, but in a major way.

The pavement is no longer able to absorb and transmit the wheel loading through the fabric of the road without causing fairly rapid further deterioration of the road pavement.

The layers making up the road pavement have failed for various reasons, but one of the most common is poor drainage, either by :-

Lack of adequate drainage provision in the original road pavement design,

Lack of maintenance of the drainage so that it no longer functions in a correct manner and the water table has risen thus weakening the road pavement,

Failure of the impervious nature of the surface course (wearing course), or the binder course (basecourse) where the surface course is a porous textured material, thus allowing the passage of surface water in to the road pavement matrix.

DRAINAGE DISTRICTS

Drainage districts occur in England and Wales, varying in size from a few hundred acres to over 100,000 acres (400 km2), all in low lying areas of the country where flood risk management and land drainage are sensitive issues.

Most drainage districts are administered by an internal drainage board(IDB), which are single purpose local drainage authorities, dealing with the drainage and water level management of clean water only.

Each drainage district has a defined area, and the IDB only has powers to deal with matters affecting that area.

Drainage districts were also used in Chicago before the Tunnel and Reservoir Plan created underground reservoirs to hold storm water.

INTERNAL DRAINAGE BOARD

An internal drainage board (IDB) is a type of operating authority which is established in areas of special drainage need in England and Wales with permissive powers to undertake work to secure clean water drainage and water level management within drainage districts. The area of an IDB is not determined by county or metropolitan council boundaries, but by water catchment areas within a given region. IDBs are geographically concentrated in the Broads, Fens in East Anglia and Lincolnshire,Somerset Levels and Yorkshire. In comparison with public bodies in other countries, IDBs are most similar to the Waterschappen of the Netherlands, Consorzi di bonifica e irrigazione of taly, wateringenof Flanders and Northern France and watershed districts of Minnesota, USA.

Much of their work involves the maintenance of rivers, drainage channels (rhynes), ordinary watercourses, pumping stations and other critical infrastructure, facilitating drainage of new developments, the ecological conservation and enhancement of watercourses, monitoring and advising on planning applications and making sure that any development is carried out in line with legislation. IDBs are not responsible for watercourses designated as main rivers within their drainage districts; the supervision of these watercourses is undertaken by the Environment Agency.

History

Internal drainage boards date back to 1252; however, the majority of today's IDBs were established by national government following the passing of the Land Drainage Act 1930 and predominantly operate under the Land Drainage Act 1991 under which, an IDB is required to exercise a general supervision over all matters relating to water level management of land within its district.

Some IDBs may also have other duties, powers and responsibilities under specific legislation for the district (for instance the Middle Level Commissioners are also a navigation authority). IDBs are responsible to Defra from whom all legislation/ regulations affecting them are issued. The work of an IDB is closely linked with that of the Environment Agency which has a range of functions providing a supervisory role over them.

Internal Drainage Boards of England and Wales

Fundamental Role

The fundamental role of an internal drainage board is to manage the water level within its district. The majority of lowland rivers and watercourses have been heavily modified by man or are totally artificial channels. All are engineered structures designed and constructed for the primary function of conveying surplus run-off to their outfall efficiently and safely, managing water levels to sustain a multitude of land functions. As with any engineered structure it must be maintained in order to function at or near its design capacity. Annual or bi-annual vegetation clearance and periodic de-silting (dredging) of these rivers and watercourses is therefore an essential component of the whole life cycle of these watercourses.

Accommodating sustainability within the design and maintenance process for lowland rivers and watercourses has to address three essential elements:

- year round conveyance of flows,
- storage of flood peaks, and
- retention and protection of flora and fauna dependent on or resident in the water corridor.

Many IDBs are redesigning watercourses to create a two-stage or bermed channel. These have been extensively created in the Lindsey Marsh Drainage Board area of East Lincolnshire to accommodate the three elements of lowland watercourse sustainability.

Berms are created at or near to the normal retained water level in the system. It is sometimes replanted with vegetation removed from the watercourse prior to improvement works but is often left to re-colonise naturally. In all cases this additional part of the channel profile allows for enhanced environmental value to develop.

The area created above the berm also provides additional flood storage capacity whilst the low level channel can be maintained in such a manner that design conveyance conditions are achieved and flood risk controlled.

By widening the channel and the berm, the berm can be safely used as access for machinery carrying out channel maintenance. Whilst, in-channel habitat that develops can be retained for a much longer period during the summer months, flood storage is provided for rare or extreme events and a buffer zone between the channel and any adjacent land use is created.

The timing of vegetation clearance works is essential to striking a sustainable balance in lowland watercourses. The Conveyance Estimating System (CES) is a modelling tool developed through a Defra/Environment Agency research collaboration. IDBs use CES to estimate the seasonal variation of conveyance owing to vegetation growth and other physical parameters which they use to assess the impact of varying the timing of vegetation clearance operations.

This is critical during the Spring and early summer, the prime nesting season for aquatic birds, the breeding season for many protected mammal species such as water voles and the season when many rare species of plant life flower and seed. Many IDBs have developed vegetation control strategies in co-ordination with Natural England.

Drainage Rates

All properties within a drainage district are deemed to derive benefit from the activities of an IDB. Every property is therefore subject to a drainage rate paid annually to the IDB.

For the purposes of rating, properties are divided into:

- Agricultural land and buildings
- Other land (such as domestic houses, factories, shops etc).

Occupiers of all "other land" pay Council Tax or Non-Domestic Rates to the local authority who then are charged by the board. This charge is called the "Special Levy". The board, therefore, only demands drainage rates direct on agricultural land and buildings. The basis of this is that each property has been allotted an "annual value" which were last revised in the early 1990s. The annual value is an amount equal to the yearly rent, or the rent that might be reasonably expected if let on a tenancy from year to year commencing 1 April 1988. The annual value remains the same from year to year. Each year the board lays a rate "in the £" to meet its estimated expenditure. This is multiplied by the annual value to produce the amount of drainage rate due on each property.

Drainage districts are local bodies formed for the purpose of draining, ditching, and improving land for agricultural and sanitary purposes. They are authorized to build and maintain drains and levees, to sue all necessary private land within their corporate bodies for that purpose, and to tax land within their boundaries as necessary.

The Illinois Constitution of 1870 authorized the General Assembly to pass laws giving landowners drainage rights, including the use of adjoining land for ditching purposes. As a result, a comprehensive drainage law was passed in 1871. The law set up legal procedures for local citizens to petition the county courts for drainage works, assessing and collecting the costs of the drainage construction from the owners of the lands to be benefited by the work, and compensating the owners of land which would be entered for ditching purposes. The county courts were given authority to appoint three drainage commissioners; township commissioners of highways could also serve as drainage commissioners.

The 1871 law was found unconstitutional; as a result the Illinois Constitution was amended, making drainage commissioners the heads of corporate drainage districts and giving these districts constitutional authority to levy property taxes. Two separate and coequal Illinois drainage laws were passed in 1879. One, the "Levee Law," repeated the procedures of the 1871 law, with added procedures for legal appeal by landowners dissatisfied with their assessments; the second, the

"Drainage District Law," made the township highway commissioners the township drainage district commissioners. For non-township-organized counties, the county commissioners served as drainage commissioners. For districts extending over three or more townships, districts were set up and temporary commissioners appointed by county courts, similarly to the "Levee Law," but if the drainage district contained more than 15 landowners, the owners elected the three commissioners. Both laws gave commissioners three-year terms. From 1871 to 1885, Illinois law required that all drainage commissions have three members. In 1885, drainage districts in which drainage construction had been completed were authorized to appoint only one commissioner. Although Illinois drainage law was recodified in 1955, the responsibilities of drainage commissioners have largely remained unchanged since 1871. From 1871 to the present, the commissioners have been required to file annual financial reports with the county or circuit courts.

Purpose of Drainage Districts

Drainage districts were established for drainage of surface waters from agricultural and other lands for the protection of said lands from overflow when said protection is a public benefit or is conducive to public health, convenience, and welfare (Section 468.2, State Code Of Iowa).

Districts are established by the Drainage District Trustees at the request of the land owners within the proposed district (Section 468.6-468.8). Petitions and actions to establish are kept in the minute books in the county courthouse.

Under the Code of Iowa, Chapter 468 the County Board Of Supervisor's act as Drainage District Trustees in all District matters. The landowners of a particular district may, if they wish, elect their own trustees and maintain the district themselves.

Requests for repairs, complaints, problems, or questions should be directed to Drainage District Clerk Paul Ketelsen at 563-659-8149 located in the Clinton County DeWitt Annex at 329 E. 11th St. DeWitt.

The trustees are required by the Code Of Iowa (Section 468.126) to maintain all drainage districts at their original capacity. Notice of repairs is only required when the cost will exceed $15,000 or 75% of the original assessed value of the district, whichever is greater.

Land owners in the district, not the County, own and pay for all maintenance and repairs to that district. A contractor hired by the trustees will do the work and bill the cost to the district.

Assessments are made as necessary to pay for engineering costs, improvements, and repairs within a given drainage district.

Assessments or classification of land in a drainage district is based on the benefit that land is seen to receive from being in the district.

- The original assessed value of any parcel of land within a district was set when that district was established and is the basis for all assessments unless the district is reclassified by the trustees (Section 468.49).

- Assessments represent a per centage of the original assessment.
- Properties near the bottom of the district or adjacent to a district ditch or tile line will normally pay more than properties at the top of the district or that are not close to a district tile. District maps show where property is located within the district in relation to the district ditches and tile. Drainage ditch and tile maps are stored in the Drainage District Clerk's Office. Copies can be made upon request.

Drainage districts are not levied on a regular basis. Drainage districts are levied when district funds drop to zero. When a district is levied the trustees set the per centage such that a surplus remains after all bills are paid. The district will not be levied again until its funds are depleted which depends on the amount of work required in the district. Some districts are levied frequently while some may go 50 years or more without a levy.

Self-managed Irrigation and Drainage Districts

Self-financing Irrigation and Drainage Districts (SIDDs) in China are a type of PIM. To promote long-run sustainability, SIDDs emphasize self-financing. However, the basic principle behind both PIM and the SIDDs is that farmers should participate and have some management control in decisions which affect their future and livelihood. The SIDD concept or system has been developed in China to match Chinese conditions and deal with many of the key issues facing Chinese irrigation management. The SIDD system is also designed to incorporate and implement Chinese policies and reforms, like "market oriented" development, the "user pays" principle, treatment of water as a commodity, water charges to cover costs, water charges according to volume of water used to encourage efficient use, and direct participation and self-management by the water users who contribute to the investments.

To date, some 490 WUAs and 38 WSCs or WSOs (Water Supply Organizations) have been established in 10 provinces and three large municipalities including Beijing on a pilot basis, supported under six ongoing or proposed World Bank-financed projects. The results of these SIDD pilot projects have generally been positive, especially for the WUAs, with increased production, reduced water waste, increased water charge collections, reduction or elimination of water conflicts among farmers, and increased "ownership," participation and contributions by farmers.

A key challenge for the future of SIDD and WUA development will be how to maintain WUA quality as they spread to wider areas, when few Water Resources staff have training in and experience with the participatory concepts on which their work with WUAs are based. A second key challenge will be how to overcome the internal government bureaucratic and institutional constraints faced by the WSCs or any type of WSO reform, which is essential to reform main system management. The need for this reform will become increasingly important in order to save water and provide high quality water supply to users, as the WUAs spread.

5

Canals

Canals are man-made channels for water. There are two types of canal:

1. Aqueducts: water supply canals that are used for the conveyance and delivery of potable water for human consumption, municipal uses, and agriculture irrigation. Rills and Acequias are small versions.
2. Waterways: navigable transportation canals used for carrying ships and boats shipping goods and conveying people.

- Those connected to existing lakes, rivers, or oceans. Included are inter-basin canals, such as the Suez Canal, Erie Canal, and the Panama Canal.
- Those connected in a city network: such as the Canal Grande and others of Venice Italy; the gracht of Amsterdam, and the waterways of Bangkok Thailand.

Types of artificial waterways

Canals are created in one of three ways, or a combination of the three, depending on available water and available path:

- A canal can be created where no stream presently exists. The body of the canal is either dug or the sides are created by piling dirt, stone, concrete, or other building materials. The water for the canal must be provided from an external source like other streams or reservoirs. Examples include canals that connect valleys over a higher body of land, like Canal du Midi and Canal de Briare.
- A stream can be canalized to make its navigable path more predictable and easier to maneuver. Canalization modifies the stream to more safely carry traffic by controlling the flow of the stream with dredging, damming, and modifying its path. Examples include Basse Saône, Canal de Mines de Fer de la Moselle, and Aisne River. Riparian zone restoration may be required.
- When a stream is too difficult to modify with canalization, a second stream can be created next to the existing stream. This is called a lateral canal. The existing stream usually acts as the water source and its banks provide a path for the new

body. Examples include the Chesapeake and Ohio Canal, Canal latéral à la Loire, Garonne Lateral Canal, and Canal latéral à l'Aisne.

Smaller transportation canals can carry barges or narrowboats, while ship canals allow seagoing ships to travel to an inland port (e.g.: Manchester Ship Canal), or from one sea or ocean to another (e.g.: Caledonian Canal, Panama Canal).

Features

At their simplest, canals consist of a trench filled with water. Depending on the stratum the canal passes through, it may be necessary to line the cut with some form of watertight material such as clay or concrete. When this is done with clay this is known as puddling.

Canals need to be level, and while small irregularities in the lie of the land can be dealt with through cuttings and embankments, for larger deviations, other approaches have been adopted. The most common is the pound lock which consists of a chamber within which the water level can be raised or lowered connecting either two pieces of canal at a different level or the canal with a river or the sea. When there is a hill to be climbed, flights of many locks in short succession may be used.

Prior to the development of the pound lock in 984AD in China by Chhaio Wei-Yo and later in Europe in the 15th century, either flash locks consisting of a single gate were used or ramps, sometimes equipped with rollers, were used to change level. Flash locks were only practical where there was plenty of water available.

Locks use a lot of water, so builders have adopted other approaches. These include boat lifts, such as the Falkirk wheel, which use a caisson of water in which boats float while being moved between two levels; and inclined planes where a caisson is hauled up a steep railway.

To cross a stream or road, the solution is usually to bridge with an aqueduct. To cross a wide valley (where the journey delay caused by a flight of locks at either side would be unacceptable) the centre of the valley can be spanned by an aqueduct - a famous example in Wales is the Pontcysyllte aqueduct across the valley of the River Dee.

Another option for dealing with hills is to tunnel through them. An example of this approach is the Harecastle Tunnel on the Trent and Mersey Canal. Tunnels are only practical for smaller canals.

Some canals attempted to keep changes in level down to a minimum. These canals known as contour canals would take longer winding routes, along which the land was a uniform altitude. Other generally later canals took more direct routes requiring the use of various methods to deal with the change in level.

Canals have various features to tackle the problem of water supply. In some cases such as the Suez Canal the canal is simply open to the sea. Where the canal is not at sea level a number of approaches have been adopted. Taking water from existing rivers or springs was an option in some cases, sometimes supplemented by other methods to deal with seasonal variations in flow. Where such sources were unavailable, reservoirs, either separate from the canal, or built into its course, and back pumping was used to provide the required water. In other cases water pumped from mines was used to feed the canal. In certain cases extensive "feeder canals" were built to bring water from sources located far from the canal.

Where large amounts of goods are loaded or unloaded such as at the end of a canal a canal basin may be built. This would normally be a section of water wider than the general canal. In some cases the canal basins contain wharfs and cranes to assist with movement of goods.

When a section of the canal needs to be sealed off so it can be drained for maintenance stop planks are frequently used. These consist of planks of wood placed across the canal to form a dam. They are generally placed in pre existing grooves in the canal bank. On more modern canals, "guard locks" or gates were sometimes placed to allow a section of canal to be quickly closed off, either for maintenance, or to prevent a major loss of water due to a canal breach.

The oldest known canals were irrigation canals, built in Mesopotamia circa 4000 BC, in what is now modern day Iraq and Syria. The Indus Valley Civilization, in Pakistan and North India, (circa 2600 BC) had sophisticated irrigation and storage systems developed, including the reservoirs built at Girnar in 3000 BC. In Egypt, canals date back at least to the time of Pepi I Meryre (reigned 2332-2283 BC), who ordered a canal built to bypass the cataract on the Nile near Aswan.

In ancient China, large canals for river transport were established as far back as the Warring States (481-221 BC), the longest one of that period being the Hong Gou (Canal of the Wild Geese), which according to the ancient historian Sima Qian connected the old states of Song, Zhang, Chen, Cai, Cao, and Wei.

By far the longest canal was the Grand Canal of China, still the longest canal in the world today. It is 1,794 kilometres (1,115 mi) long and was built to carry the Emperor Yang Guang between Beijing and Hangzhou. The project began in 605 and was completed in 609, although much of the work combined older canals, the oldest section of the canal existing since at least 486 BC. Even in its narrowest urban sections it is rarely less than 30 metres (98 ft) wide. Greek engineers were the first to use canal locks, by which they regulated the water flow in the Ancient Suez Canal as early as the 3rd century BC.

Canals in the Middle Ages

In the Middle Ages, water transport was cheaper and faster than transport overland. This was because roads were unpaved and in poor condition and greater amounts could be transported by ship. The first artificial canal in Christian Europe was the Fossa Carolina built at the end of the 8th Century under personal supervision of Charlemagne. More lasting and of more economic impact were canals like the Naviglio Grande built between 1127 and 1257, the most important of the lombard "navigli", Later, canals were built in the Netherlands and Flanders to drain the polders and assist the transportation of goods.

Canal building was revived in this age because of commercial expansion from the 12th century AD. River navigations were improved progressively by the use of single, or flash locks. Taking boats through these used large amounts of water leading to conflicts with watermill owners and to correct this, the pound or chamber lock first appeared, in 10th century AD in China and in Europe in 1373 in Vreeswijk, Netherlands. Another important development was the mitre gate which was probably introduced in Italy by Bertola da Novate in the sixteenth century. This allowed wider gates and also removed the height restriction of guillotine locks.

To break out of the limitations caused by river valleys, the first summit level canals were developed with the Grand Canal of China in 581-617 AD whilst in Europe the first, also using single locks, was the Stecknitz Canal in Germany in 1398. The first to use pound locks was the Briare Canal connecting the Loire and Seine (1642), followed by the more ambitious Canal du Midi (1683) connecting the Atlantic to the Mediterranean. This included a staircase of 8 locks at Béziers, a 157 metres (515 ft) tunnel and three major aqueducts.

Canal building progressed steadily in Germany in the 17th and 18th centuries with three great rivers, the Elbe, Oder and Weser being linked by canals. In post-Roman Britain, the first canal built appears to have been the Exeter Canal, which opened in 1563. The oldest canal built for industrial purposes in North America is Mother Brook in Dedham, MA. It was constructed in 1639 to provide water power for mills. In Russia, the Volga-Baltic Waterway, a nationwide canal system connecting the Baltic and Caspian seas via the Neva and Volga rivers, was opened in 1718.

Industrial revolution

Canals were important for industrial development. The greatest stimulus to canal system building came from the Industrial Revolution with its need for cheap transport of unprecedented quantities of raw materials and manufactured items.

In Europe, particularly Britain and Ireland, and then in the young United States and the Canadian colonies, inland canals preceded the development of railroads during the earliest phase of the Industrial Revolution. The opening of the Sankey Canal in 1757, followed by the Bridgewater Canal in 1761, which halved the price of coal in Liverpool and Manchester, respectively, triggered a period of "canal mania" in Britain so that between 1760 and 1820 over one hundred canals were built.

The Blackstone Canal in Massachusetts and Rhode Island fulfilled a similar role in the early industrial revolution between 1828-48. The Blackstone Valley was a major contributor of the American Industrial Revolution where Samuel Slater built his first mill.

In addition to their transportation purposes, parts of the United States, particularly in the Northeast, had enough fast-flowing rivers that water power was the primary means of powering factories (usually textile mills) until after the American Civil War. For example, Lowell, Massachusetts, considered to be "The Cradle of the

American Industrial Revolution," has 6 miles (9.7 km) of canals, built from around 1790 to 1850, that provided water power and a means of transportation for the city. The output of the system is estimated at 10,000 horsepower. Other cities with extensive power canal systems include Lawrence, Massachusetts, Holyoke, Massachusetts, Manchester, New Hampshire, and Augusta, Georgia.

The 19th century

Competition from the railway network from the 1830s, and in the 20th century the roads, made the smaller canals obsolete for most commercial transportation, and many of the British canals fell into decay. Only the Manchester Ship Canal and the Aire and Calder Canal bucked this trend. Yet in other countries canals grew in size as construction techniques improved. During the 19th century in the US, the length of canals grew from 100 miles (161 km) to over 4,000, with a complex network making the Great Lakes navigable, in conjunction with Canada, although some canals were later drained and used as railroad rights-of-way.

In the United States, navigable canals reached into isolated areas and brought them in touch with the world beyond. By 1825 the Erie Canal, 363 miles (584 km) long with 82 locks, opened up a connection from the populated Northeast to the Great Lakes. Settlers flooded into regions serviced by such canals, since access to markets was available. The Erie Canal (as well as other canals) was instrumental in lowering the differences in commodity prices between these various markets across America.

The canals caused price convergence between different regions because of their reduction in transportation costs, which allowed Americans to ship and buy goods from farther distances for much lower prices compared to before. Ohio built many miles of canal, Indiana had working canals for a few decades, and the Illinois and Michigan Canal connected the Great Lakes to the Mississippi River system until replaced by a channelized river waterway.

Three major canals with very different purposes were built in what is now Canada. The first Welland Canal, which opened in 1829 between Lake Ontario and Lake Erie, bypassing Niagara Falls and the Lachine Canal (1825) which allowed ships to skirt the nearly impassable rapids on the St. Lawrence River at Montreal were built for commerce.

The Rideau Canal, completed in 1832, connects Ottawa, on the Ottawa River to Kingston, Ontario on Lake Ontario. The Rideau Canal was built as a result of the War of 1812 to provide military transportation between the British colonies of Upper Canada and Lower Canada as an alternative to part of the St. Lawrence River which was susceptible to blockade by the United States.

In France, a steady linking of all the river systems-Rhine, Rhône, Saône and Seine-and the North Sea was boosted in 1879 by the establishment of the Freycinet gauge which specified the minimum size of locks so that canal traffic doubled in the first decades of the 20th century.

Many notable sea canals were completed in this period, starting with the Suez Canal (1869), and the Kiel Canal (1897), which carries tonnage many times that of most other canals, though the Panama Canal was not opened until 1914.

In the 19th century, a number of canals were built in Japan including the Biwako canal and the Tone canal. These canals were partially built with the help of engineers from the Netherlands and other countries.

Modern uses

Large scale ship canals such as the Panama Canal and Suez Canal continue to operate for cargo transportation; as do European barge canals. Due to globalization, they are becoming increasingly important, resulting in expansion projects such as the Panama Canal expansion project.

The narrow early industrial canals, however, have ceased to carry significant amounts of trade and many have been abandoned to navigation, but may still be used as a system for transportation of untreated water. In some cases railways have been built along the canal route, an example being the Croydon Canal.

A movement that began in Britain and France to use the early industrial canals for pleasure boats, such as hotel barges, has spurred rehabilitation of stretches of historic canals. In some cases abandoned canals such as the Kennet and Avon Canal have been restored and are now used by pleasure boaters. In Britain canalside housing has also proven popular in recent years.

The Seine-Nord Europe Canal is being developed into a major transportation waterway, linking France with Belgium, Germany and the Netherlands.

Canals have found another use in the 21st century, as easements for the installation of fibre optic telecommunications network cabling avoiding having them buried in roadways while facilitating access and reducing the hazard of being damaged from digging equipment.

Canals are still used to provide water for agriculture. An extensive canal system exists within the Imperial Valley in the Southern California desert to provide irrigation to agriculture within the area.

Canals are so deeply identified with Venice that many canal cities have been nicknamed "the Venice of…". The city is built on marshy islands, with wooden piles supporting the buildings, so that the land is man-made rather than the waterways. The islands have a long history of settlement; by the 12th century, Venice was a powerful city state.

Amsterdam was built in a similar way, with buildings on wooden piles. It became a city around 1300.Other cities with extensive canal networks include: Delft, Haarlem and Leiden in the Netherlands, Brugge in Flanders, Belgium, Birmingham in England - which has 35 miles of canals, to Venice's 26 miles -, Saint Petersburg in Russia, Hamburg in Germany, Berlin in Germany , Fort Lauderdale, Florida, and Cape Coral, Florida in the United States.

Liverpool Maritime Mercantile City is a UNESCO World Heritage Site near the centre of Liverpool, England, where a system of intertwining waterways and docks now being developed for mainly residential and leisure use.

Canal Estates (commonly known as bayous) are a form of subdivision popular in cities like Miami, Florida, Texas City, Texas and the Gold Coast, Queensland; the Gold Coast has over 700 km of residential canals. Wetlands are difficult areas upon which to build housing estates, so dredging part of the wetland down to a navigable channel provides fill to build up another part of the wetland above the flood level for houses. Land is built up in a finger pattern that provides a suburban street layout of waterfront housing blocks.

Inland canals have often had boats specifically built for them. An example of this is the British narrowboat, which is up to 72 feet (21.95 m) long and 7 feet (2.13 m) wide and was primarily built for British Midland canals. In this case the limiting factor was the size of the locks. This is also the limiting factor on the Panama canal where Panamax ships are limited to a length of 294.1 m (965 ft) and a width of 32.3 m (106 ft).

For the lockless Suez Canal the limiting factor for Suezmax ships is generally draft, which is limited to 16 m (52.5 ft). At the other end of the scale, tub-boat canals such as the Bude Canal were limited to boats of under 10 tons for much of their length due to the capacity of their inclined planes or boat lifts. Most canals have a limit on height imposed either by bridges or tunnels.

RIDGE

A ridge is a geological feature that features a chain of mountains or hills that are of a continuous elevated crest for some distance. Ridges are usually termed hills or mountains as well, depending on size. There are several main types of ridges:

- Dendritic ridge: In typical dissected plateau terrain, the stream drainage valleys will leave intervening ridges. These are by far the most common ridges. These ridges usually represent slightly more erosion resistant rock, but not always - they often remain because there were more joints where the valleys formed, or other chance occurrences. This type of ridge is generally somewhat random in orientation, often changing direction frequently, often with knobs at intervals on the ridge top.
- Stratigraphic ridge: In places such as the Ridge-and-valley Appalachians, long, even, straight ridges are formed because they are the uneroded remaining edges of the more resistant strata that were folded laterally. Similar ridges have formed in places such as the Black Hills, where the ridges form concentric circles around the igneous core. Sometimes these ridges are called "hogback ridges".
- Oceanic spreading ridge: In tectonic spreading zones around the world, such as at the Mid-Atlantic Ridge, the volcanic activity forming new plate boundary forms volcanic ridges at the spreading zone. Isostatic settling and erosion gradually reduce the elevations moving away from the zone.

- Crater ridges: Large meteorite strikes typically form large impact craters bordered by circular ridges.
- Volcanic caldera ridges: Large volcanoes often leave collapsed central calderas that are bordered by circular ridges.
- Fault ridges: Faults often form escarpments. Sometimes the tops of the escarpments form not plateaus, but slope back so that the edges of the escarpments form ridges.
- Dune ridges: In areas of large-scale dune activity, certain types of dunes result in sand ridges.
- Moraines and eskers: Glacial activity may leave ridges in the form of moraines and eskers. An arête is a thin ridge of rock that is formed by glacial erosion.
- Volcanic subglacial ridges: Many subglacial volcanoes create ridgelike formations when lava erupts through a thick glacier or ice sheet.

TECTONIC UPLIFT

Tectonic uplift is a geological process most often caused by plate tectonics which increases elevation. The opposite of uplift is subsidence, which results in a decrease in elevation. Uplift may be orogenic or isostatic. Orogenic uplift is the result of tectonic-plate collisions and results in mountain ranges or a more modest uplift over a large region. The Himalayas were (and are still being) formed by the collision of two continental plates, the Indian and Eurasian Plates. This ongoing collision produced the Tibetan Plateau as well as the Himalayas and associated ranges.

Ozark Plateau is a broad uplifted area which resulted from the Ouachita Orogeny to the south during the Permian Period. Another related uplift is the Llano Uplift in Texas, a geographical location named after its uplift features. The Colorado Plateau with its spectacular scenic canyons, the Grand Canyon, is also the result of broad tectonic uplift followed by river erosion.

Isostatic uplift

Isostatic uplift includes the gradual uplift following rapid erosional removal of material from a mountain range. The land rises as a result of the removal of the weight. Another example of isostatic uplift is post-glacial rebound following the melting of continental glaciers and ice sheets. The Hudson Bay region of Canada and the Great Lakes of Canada and the United States are currently undergoing gradual rebound as a result of the melting of the ice sheets 10,000 years ago.

Coral islands

In a few cases, tectonic uplift can be seen in the cases of coral islands. This is evidenced by the presence of various oceanic islands composed entirely of coral, which otherwise appear to be high islands (i.e., islands of volcanic origin). Examples of such islands are found in the Pacific, notably the three phosphate islets, Nauru,

Makatea, and Banaba as well as Maré and Lifou in New-Caledonia, Fatu Huku in the Marquesas Islands and Henderson Island in the Pitcairn Islands. The uplift of these islands is the result of the movement of oceanic tectonic plates. Sunken islands or guyots with their coral reefs are the result of crustal subsidence as the oceanic plate carries the islands to deeper or lower oceanic crust areas.

Waterlogging

It refers to the saturation of soil with water. Soil may be regarded as waterlogged when the water table of the groundwater is too high to conveniently permit an anticipated activity, like agriculture.

In agriculture, various crops need air (specifically, oxygen) to a greater or lesser depth in the soil. Waterlogging of the soil stops air getting in. How near the water table must be to the surface for the ground to be classed as waterlogged, varies with the purpose in view. A crop's demand for freedom from waterlogging may vary between seasons of the year, as with the growing of rice (Oryza sativa).

In irrigated agricultural land, waterlogging is often accompanied by soil salinity as waterlogged soils prevent leaching of the salts imported by the irrigation water.

From a gardening point of view, waterlogging is the process whereby the soil blocks off all water and is so hard it stops air getting in and it stops oxygen from getting in.

Soil salinity control

Soil salinity control relates to controlling the problem of soil salinity and reclaiming salinized agricultural land. The aim of soil salinity control is to prevent soil degradation by salinization and reclaim already salty (saline) soils. Soil reclamation is also called soil improvement, rehabilitation, remediation, recuperation, or amelioration.

The primary man-made cause of salinization is irrigation. River water or groundwater used in irrigation contains salts, which remain behind in the soil after the water has evaporated.

The primary method of controlling soil salinity is to permit 10-20% of the irrigation water to leach the soil, be drained and discharged through an appropriate drainage system. The salt concentration of the drainage water is normally 5 to 10 times higher than that of the irrigation water, thus salt export matches salt import and it will not accumulate.

Salty (saline) soils are soils that have a high salt content. The predominant salt is normally sodium chloride (NaCl, "table salt"). Saline soils are therefore also sodic soils but there may be sodic soils that are not saline, but alkaline.

Salty soils are a common feature and an environmental problem in irrigated lands in arid and semi-arid regions. They have poor or little crop production. The problems are often associated with high water tables, caused by a lack of natural subsurface drainage to the underground. Poor subsurface drainage may be caused by insufficient transport capacity of the aquifer or because water cannot exit the aquifer for instance, if it is situated in a topographical depression.

Worldwide, the major factor in the development of saline soils is a lack of precipitation. Most naturally saline soils are found in (semi)arid regions and climates of the globe.

Primary cause

The primary cause of man-made salinization is the salt brought in with irrigation water. All irrigation water derived from rivers or groundwater, however 'sweet', contains salts that remain behind in the soil after the water has evaporated, which may create the environmental issue.

For example, assuming irrigation water with a low salt concentration of 0.3 g/l (equal to 0.3 kg/m^3 corresponding to an electric conductivity of about 0.5 dS/m) and a modest annual supply of irrigation water of 10,000 m^3/ha (almost 3 mm/day) brings 3,000 kg salt/ha each year. In the absence of sufficient natural drainage (as in waterlogged soils) and without a proper leaching and drainage program to remove salts, this would lead to a high soil salinity and reduced crop yields in the long run.

Much of the water used in irrigation has a higher salt content than in this example, which is compounded by that fact that many irrigation projects use a far greater annual supply of water. Sugar cane, for example, needs about 20000 m3/ha of water per year. As a result, irrigated areas often receive more than 3,000 kg/ha of salt per year and some receive as much as 10,000 kg/ha/year.

Secondary cause

The secondary cause of salinization is waterlogging in irrigated land.

Irrigation causes changes to the natural water balance of irrigated lands. Large quantities of water in irrigation projects are not consumed by plants and must go somewhere. In irrigation projects it is impossible to achieve 100% irrigation efficiency where all the irrigation water is consumed by the plants. The maximum attainable irrigation efficiency is about 70% but usually it is less than 60%. This means that minimum 30%, but usually more than 40% of the irrigation water is not evaporated and it must go somewhere.

Most of the water lost this way is stored underground which can change the original hydrology of local aquifers considerably. Many aquifers cannot absorb and transport these quantities of water and, the water table rises leading to water logging.

Waterlogging causes three problems:

- the shallow water table and lack of oxygenation of the root zone reduces the yield of most crops
- it leads to an accumulation of salts brought in with the irrigation water as their removal through the aquifer is blocked
- with the upward seepage of groundwater more salts are brought into the soil and the salination is aggravated

Aquifer conditions in irrigated land and the groundwater flow have an important role in soil salinization, as illustrated here :

- Illustration of the influence of aquifer conditions on soil salinization in irrigated land
- Soil salinization in the lower parts of undulating land with a good aquifer
- Soil salinization in the unirrigated parts of flat land with a good aquifer
- Soil salinization in irrigated flat land without an aquifer

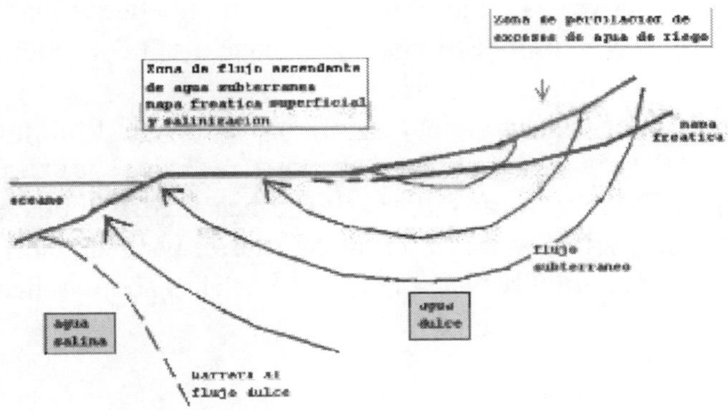

Soil salinization in a coastal delta from irrigation higher up

Salt affected area

Normally, the salinization of agricultural land affects a considerable area of irrigation projects, on the order of 20 to 30%. When the agriculture in such a fraction of the land is abandoned, a new salt and water balance is attained, a new equilibrium is reached, and the situation becomes stable.

In India alone, thousands of square kilometres have been severely salinized. China and Pakistan do not lag much behind (perhaps China has even more salt affected land than India). A regional distribution of the 3,230,000 km² of saline land world wide is shown in the following table derived from the FAO/UNESCO Soil Map of the World.

Spatial variation

Although the principles of the processes of salinization are fairly easy to understand, it is more difficult to explain why certain parts of the land suffer from the problems and other parts do not, or to predict accurately which part of the land will fall victim. The main reason for this is the variation of natural conditions in time and space, the usually uneven distribution of the irrigation water, and the seasonal or yearly changes of agricultural practices.

Only in lands with undulating topography the explanation and prediction is pretty simple: the depressional areas will degrade the most.

The preparation of salt and water balances for distinguishable sub-areas in the irrigation project, or the use of agro-hydro-salinity models, can be helpful in explaining or predicting the extent and severity of the problems.

Diagnosis

Measurement

Soil salinity is measured as the salt concentration of the soil solution in tems of g/l or electric conductivity (EC) in dS/m. The relation between these two units is about 5/3 : y g/l => 5y/3 dS/m. Sea water may have a salt concentration of 30 g/l (3%) and an EC of 50 dS/m.

The standard for the determination of soil salinity is from an extract of a saturated paste of the soil, and the EC is then written as ECe. The extract is obtained by centrifugation. The salinity can more easily be measured, without centrifugation, in a 2:1 or 5:1 water:soil mixture (in terms of g water per g dry soil) than from a saturated paste. The relation between ECe and EC2:1 is about 4, hence : ECe = 4 EC1:2.

Classification

Soils are considered saline when the ECe > 4. When 4 < ECe < 8, the soil is called slightly saline, when 8 < ECe < 16 it is called (moderately) saline, and when ECe > 16 severely saline.

Crop tolerance

Sensitive crops loose their vigor already in slightly saline soils, most crops are negatively affected by (moderately) saline soils, and only salinity resistant crops thrive in severely saline soils. The University of Wyoming and the Government of Alberta report data on the salt tolerance of plants.

PRINCIPLES OF SALINITY CONTROL

Drainage is the primary method of controlling soil salinity. The system should permit a small fraction of the irrigation water (about 10 to 20 percent, the drainage or leaching fraction) to be drained and discharged out of the irrigation project.

In irrigated areas where salinity is stable, the salt concentration of the drainage water is normally 5 to 10 times higher than that of the irrigation water. Salt export matches salt import and salt will not accumulate.

When reclaiming already salinized soils, the salt concentration of the drainage water will initially be much higher than that of the irrigation water (for example 50 times higher). Salt export will greatly exceed salt import, so that with the same drainage fraction a rapid desalinization occurs. After one or two years, the soil salinity is decreased so much, that the salinity of the drainage water has come down to a normal value and a new, favorable, equilibrium is reached.

In regions with pronounced dry and wet seasons, the drainage system may be operated in the wet season only, and closed during the dry season. This practice of checked or controlled drainage saves irrigation water.

The discharge of salty drainage water may pose environmental problems to downstream areas. The environmental hazards must be considered very carefully and, if necessary mitigating measures must be taken. If possible, the drainage must be limited to wet seasons only, when the salty effluent inflicts the least harm.

Drainage systems

The drainage system designed to evacuate salty water also lowers the water table. To reduce the cost of the system, the lowering must be reduced to a minimum. The highest permissible level of the water table (or the shallowest permissible depth) depends on the irrigation and agricultural practices and kind of crops.

In many cases a seasonal average water table depth of 0.6 to 0.8 m is deep enough. This means that the water table may occasionally be less than 0.6 m (say 0.2 m just after an irrigation or a rain storm). This automatically implies that, in other occasions, the water table will be deeper than 0.8 m (say 1.2 m). The fluctuation of the water table helps in the breathing function of the soil while the expulsion of carbon dioxide (CO2) produced by the plant roots and the inhalation of fresh oxygen (O2) is promoted.

The establishing of a not too deep water table offers the additional advantage that excessive field irrigation is discouraged, as the crop yield would be negatively affected by the resulting elevated water table, and irrigation water may be saved.

The statements made above on the optimum depth of the watertable are very general, because in some instances the required water table may be still shallower than indicated (for example in rice paddies), while in other instances it must be considerably deeper (for example in some orchards). The establishment of the optimum depth of the water table is in the realm of agricultural drainage criteria.

Soil leaching

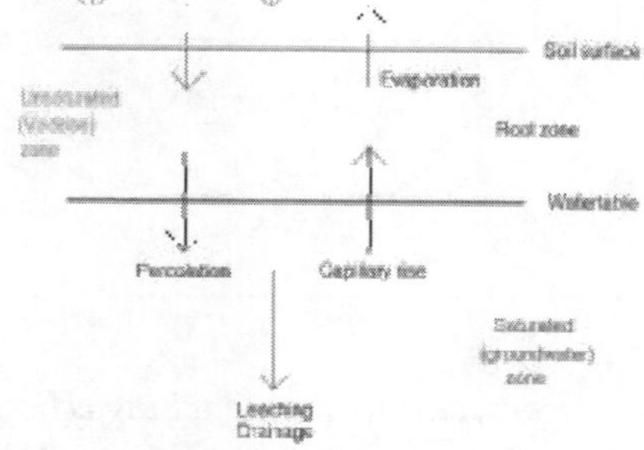

Water balance factors in the soil

The unsaturated zone or vadose zone of the soil below the soil surface and the watertable is subject to four main hydrological inflow and outflow factors :

- Infiltration of rain and irrigation water (Irr) into the soil through the soil surface (Inf) :
- Inf = Rain + Irr
- Evaporation of soil water through plants and directly into the air through the soil surface (Evap)
- Percolation of water from the unsaturated zone soil into the groundwater through the watertable (Perc)
- Capillary rise of groundwater moving by capillary suction forces into the unsaturated zone(Cap)

In steady state (i.e. the amount of water stored in the unsaturated zone does not change in the long run) the water balance of the unsaturated zone reads: Inflow = Outflow, thus:

- Inf + Cap = Evap + Perc or :
- Irr + Rain + Cap = Evap + Perc and the salt balance is
- Irr.Ci + Cap.Cc = Evap.Fc.Ce + Perc.Cp + Ss

where Ci is the salt concentration of the irrigation water, Cc is the salt concentration of the capillary rise, equal to the salt concentration of the upper part of the groundwater body, Fc is the fraction of the total evaporation transpired by plants, Ce is the salt concentration of the water taken up by the plant roots, Cp is the salt concentration of the percolation water, and Ss is the increase of salt storage in the unsaturated soil. This assumes that the rainfall contains no salts. Only along the coast this may not be true. Further it is assumed that no runoff or surface drainage occurs.

The amount of salts removed by plants (Evap.Fc.Ce) is usually negligibly small: Evap.Fc.Ce = 0

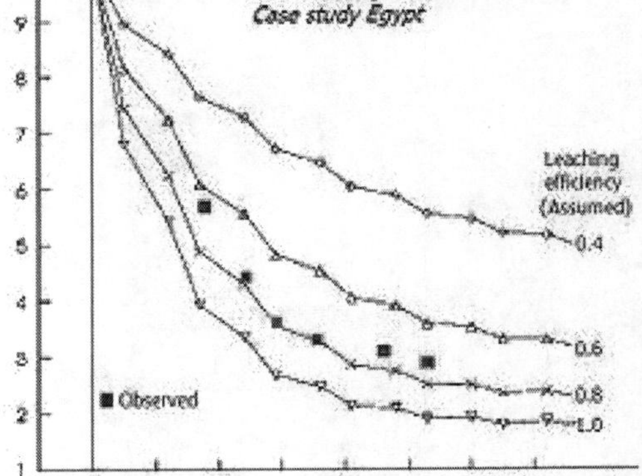

Leaching curves, calibrating leaching efficiency

The salt concentration Cp can be taken as a part of the salt concentration of the soil in the unsaturated zone (Cu) giving: Cp=Le.Cu, where Le is the leaching efficiency. The leaching efficiency is often in the order of 0.7 to 0.8, but in poorly structured, heavy clay soils it may be less. In the Leziria Grande polder in the delta of the Tagus river in Portugal it was found that the leaching efficiency was only 0.15.

Assuming that one wishes to avoid the soil salinity to increase and maintain the soil salinity Cu at a desired level Cd we have:

Ss = 0, Cu = Cd and Cp = Le.Cd. Hence the salt balance can be simplified to:

- Perc.Le.Cd = Irr.Ci + Cap.Cc

Setting the amount percolation water required to fulfill this salt balance equal to Lr (the leaching requirement) it is found that:

- Lr = (Irr.Ci + Cap.Cc) / Le.Cd .

Substituting herein Irr = Evap + Perc ? Rain ? Cap and re-arranging gives :

- Lr = [(Evap?Rain).Ci + Cap(Cc?Ci)] / (Le.Cd ? Ci)

With this the irrigation and drainage requirements for salinity control can can be computed too.

In irrigation projects in (semi)arid zones and climates it is important to check the leaching requirement, whereby the field irrigation efficiency (indicating the fraction of irrigation water percolating to the underground) is to be taken into account.

The desired soil salinity level Cd depends on the crop tolerance to salt. The University of Wyoming, USA, and the Government of Alberta, Canada, report crop tolerance data.

Strip cropping: an alternative

In irrigated lands with scarce water resources suffering from drainage (high water table) and soil salinity problems, strip cropping is sometimes practiced with strips of land where every other strip is irrigated while the strips in between are left permanently fallow.

Owing to the water application in the irrigated strips they have a higher watertable which induces flow of groundwater to the unirrigated strips. This flow functions as subsurface drainage for the irrigated strips, whereby the water table is maintained at a not-too-shallow depth, leaching of the soil is possible, and the soil salinity can be controlled at an acceptably low level.

In the unirrigated (sacrificial) strips the soil is dry and the groundwater comes up by capillary rise and evaporates leaving the salts behind, so that here the soil salinizes. Nevertheless, they can have some use for livestock, sowing salinity resistant grasses or weeds. Moreover, useful salt resistant trees can be planted like Casuarina, Eucalyptus or Atriplex, keeping in mind that the trees have deep rooting systems and the salinity of the wet subsoil is less than of the topsoil. In these ways wind erosion can be controlled. The unirrigated strips can also be used for salt harvesting.

The majority of the computer models available for water and solute transport in the soil (e.g. SWAP, DrainMod-S, UnSatChem), are based on Richard's differential equation for the movement of water in unsaturated soil in combination with a differential salinity dispersion equation.

The models require input of soil characteristics like the relation between unsaturated soil moisture content, water tension, hydraulic conductivity and dispersivity. These relations vary to a great extent from place to place and are not easy to measure. The models use short time steps and need at least a daily data base of hydrological phenomena. Altogether this makes model application to a fairly large project the job of a team of specialists with ample facilities.

Simpler models, like SaltMod, based on seasonal water and soil balances and an empirical capillary rise function, are also available. They are useful for long-term salinity predictions in relation to irrigation and drainage practices.Spatial variations owing to variations in topography can be simulated and predicted using salinity cum groundwater models, like SahysMod.

WATER RESOURCES

Water resources are sources of water that are useful or potentially useful to humans. Uses of water include agricultural, industrial, household, recreational and environmental activities. Virtually all of these human uses require fresh water.

97% of water on the Earth is salt water, and only 3% is fresh water of which slightly over two thirds is frozen in glaciers and polar ice caps. The remaining unfrozen freshwater is mainly found as groundwater, with only a small fraction present above ground or in the air.

Fresh water is a renewable resource, yet the world's supply of clean, fresh water is steadily decreasing. Water demand already exceeds supply in many parts of the world and as the world population continues to rise, so too does the water demand. Awareness of the global importance of preserving water for ecosystem services has only recently emerged as, during the 20th century, more than half the world's wetlands have been lost along with their valuable environmental services. Biodiversity-rich freshwater ecosystems are currently declining faster than marine or land ecosystems. The framework for allocating water resources to water users (where such a framework exists) is known as water rights.

Sources of fresh water

Surface water is water in a river, lake or fresh water wetland. Surface water is naturally replenished by precipitation and naturally lost through discharge to the oceans, evaporation, evapotranspiration and sub-surface seepage.

Although the only natural input to any surface water system is precipitation within its watershed, the total quantity of water in that system at any given time is also dependent on many other factors. These factors include storage capacity in lakes, wetlands and artificial reservoirs, the permeability of the soil beneath these storage bodies, the runoff characteristics of the land in the watershed, the timing of the precipitation and local evaporation rates. All of these factors also affect the proportions of water lost.

Human activities can have a large and sometimes devastating impact on these factors. Humans often increase storage capacity by constructing reservoirs and decrease it by draining wetlands. Humans often increase runoff quantities and velocities by paving areas and channelizing stream flow.

The total quantity of water available at any given time is an important consideration. Some human water users have an intermittent need for water. For example, many farms require large quantities of water in the spring, and no water at all in the winter. To supply such a farm with water, a surface water system may require a large storage capacity to collect water throughout the year and release it in a short period of time. Other users have a continuous need for water, such as a power plant that requires water for cooling. To supply such a power plant with water, a surface water system only needs enough storage capacity to fill in when average stream flow is below the power plant's need.

Nevertheless, over the long term the average rate of precipitation within a watershed is the upper bound for average consumption of natural surface water from that watershed.

Natural surface water can be augmented by importing surface water from another watershed through a canal or pipeline. It can also be artificially augmented from any of the other sources listed here, however in practice the quantities are negligible. Humans can also cause surface water to be "lost" (i.e. become unusable) through pollution.

Brazil is the country estimated to have the largest supply of fresh water in the world, followed by Russia and Canada.

Under river flow

Throughout the course of the river, the total volume of water transported downstream will often be a combination of the visible free water flow together with a substantial contribution flowing through sub-surface rocks and gravels that underlie the river and its floodplain called the hyporheic zone. For many rivers in large valleys, this unseen component of flow may greatly exceed the visible flow.

The hyporheic zone often forms a dynamic interface between surface water and true ground-water receiving water from the ground water when aquifers are fully charged and contributing water to ground-water when ground waters are depleted. This is especially significant in karst areas where pot-holes and underground rivers are common.

Ground water

Sub-surface water, or groundwater, is fresh water located in the pore space of soil and rocks. It is also water that is flowing within aquifers below the water table. Sometimes it is useful to make a distinction between sub-surface water that is closely associated with surface water and deep sub-surface water in an aquifer (sometimes called "fossil water").

Sub-surface water can be thought of in the same terms as surface water: inputs, outputs and storage. The critical difference is that due to its slow rate of turnover, sub-surface water storage is generally much larger compared to inputs than it is for surface water. This difference makes it easy for humans to use sub-surface water unsustainably for a long time without severe consequences. Nevertheless, over the long term the average rate of seepage above a sub-surface water source is the upper bound for average consumption of water from that source.

The natural input to sub-surface water is seepage from surface water. The natural outputs from sub-surface water are springs and seepage to the oceans. If the surface water source is also subject to substantial evaporation, a sub-surface water source may become saline. This situation can occur naturally under endorheic bodies of water, or artificially under irrigated farmland.

In coastal areas, human use of a sub-surface water source may cause the direction of seepage to ocean to reverse which can also cause soil salinization. Humans can also cause sub-surface water to be "lost" (i.e. become unusable) through pollution. Humans can increase the input to a sub-surface water source by building reservoirs or detention ponds.

Desalination

Desalination is an artificial process by which saline water (generally sea water) is converted to fresh water. The most common desalination processes are distillation and reverse osmosis. Desalination is currently expensive compared to most alternative sources of water, and only a very small fraction of total human use is satisfied by desalination. It is only economically practical for high-valued uses (such as household and industrial uses) in arid areas. The most extensive use is in the Persian Gulf.

Frozen water

Several schemes have been proposed to make use of icebergs as a water source, however to date this has only been done for novelty purposes. Glacier runoff is considered to be surface water.

The Himalayas, which are often called "The Roof of the World", contain some of the most extensive and rough high altitude areas on Earth as well as the greatest area of glaciers and permafrost outside of the poles. Ten of Asia's largest rivers flow from there, and more than a billion people's livelihoods depend on them. To complicate matters, temperatures are rising more rapidly here than the global average.

In Nepal the temperature has risen with 0.6 degree over the last decade, whereas the global warming has been around 0.7 over the last hundred years. Uses of fresh water can be categorized as consumptive and non-consumptive (sometimes called "renewable"). A use of water is consumptive if that water is not immediately available for another use. Losses to sub-surface seepage and evaporation are considered consumptive, as is water incorporated into a product (such as farm produce). Water that can be treated and returned as surface water, such as sewage, is generally considered non-consumptive if that water can be put to additional use.

Agricultural

It is estimated that 69% of worldwide water use is for irrigation, with 15-35% of irrigation withdrawals being unsustainable. It takes around 3,000 litres of water, converted from liquid to vapour, to produce enough food to satisfy one person's daily dietary need. This is a considerable amount, when compared to that required for drinking, which is between two and five litres. To produce food for the 6.5 billion or so people who inhabit the planet today requires the water that would fill a canal ten metres deep, 100 metres wide and 7.1 million kilometres long - that's enough to circle the globe 180 times.

Increasing water scarcity

Fifty years ago, the common perception was that water was an infinite resource. At this time, there were fewer than half the current number of people on the planet. People were not as wealthy as today, consumed fewer calories and ate less meat, so less water was needed to produce their food. They required a third of the volume of water we presently take from rivers. Today, the competition for water resources is much more intense.

This is because there are now nearly seven billion people on the planet, their consumption of water-thirsty meat and vegetables is rising, and there is increasing competition for water from industry, urbanisation and biofuel crops. In future, even more water will be needed to produce food because the Earth's population is forecast to rise to 9 billion by 2050. An additional 2.5 or 3 billion people, choosing to eat fewer cereals and more meat and vegetables could add an additional five million kilometres to the virtual canal mentioned above.

An assessment of water management in agriculture was conducted in 2007 by the International Water Management Institute in Sri Lanka to see if the world had sufficient water to provide food for its growing population. It assessed the current availability of water for agriculture on a global scale and mapped out locations suffering from water scarcity.

It found that a fifth of the world's people, more than 1.2 billion, live in areas of physical water scarcity, where there is not enough water to meet all demands. A further 1.6 billion people live in areas experiencing economic water scarcity, where the lack of investment in water or insufficient human capacity make it impossible for authorities to satisfy the demand for water.

The report found that it would be possible to produce the food required in future, but that continuation of today's food production and environmental trends would lead to crises in many parts of the world. To avoid a global water crisis, farmers will have to strive to increase productivity to meet growing demands for food, while industry and cities find ways to use water more efficiently.

In some areas of the world irrigation is necessary to grow any crop at all, in other areas it permits more profitable crops to be grown or enhances crop yield. Various irrigation methods involve different trade-offs between crop yield, water consumption and capital cost of equipment and structures. Irrigation methods such as furrow and overhead sprinkler irrigation are usually less expensive but are also typically less efficient, because much of the water evaporates, runs off or drains below the root zone.

Other irrigation methods considered to be more efficient include drip or trickle irrigation, surge irrigation, and some types of sprinkler systems where the sprinklers are operated near ground level. These types of systems, while more expensive, usually offer greater potential to minimize runoff, drainage and evaporation. Any system that is improperly managed can be wasteful, all methods have the potential for high efficiencies under suitable conditions, appropriate irrigation timing and management. Some issues that are often insufficiently considered are salinization of sub-surface water and contaminant accumulation leading to water quality declines.

As global populations grow, and as demand for food increases in a world with a fixed water supply, there are efforts under way to learn how to produce more food with less water, through improvements in irrigation methods and technologies, agricultural water management, crop types, and water monitoring. Aquaculture is a small but growing agricultural use of water. Freshwater commercial fisheries may also be considered as agricultural uses of water, but have generally been assigned a lower priority than irrigation.

Industrial

It is estimated that 22% of worldwide water use is industrial. Major industrial users include power plants, which use water for cooling or as a power source (i.e. hydroelectric plants), ore and oil refineries, which use water in chemical processes, and manufacturing plants, which use water as a solvent.

The portion of industrial water usage that is consumptive varies widely, but as a whole is lower than agricultural use. Water is used in power generation. Hydroelectricity is electricity obtained from hydropower. Hydroelectric power comes from water driving a water turbine connected to a generator. Hydroelectricity is a low-cost, non-polluting, renewable energy source. The energy is supplied by the sun. Heat from the sun evaporates water, which condenses as rain in higher altitudes, from where it flows down.

Pressurized water is used in water blasting and water jet cutters. Also, very high pressure water guns are used for precise cutting. It works very well, is relatively safe, and is not harmful to the environment. It is also used in the cooling of machinery to prevent over-heating, or prevent saw blades from over-heating.

Water is also used in many industrial processes and machines, such as the steam turbine and heat exchanger, in addition to its use as a chemical solvent. Discharge of untreated water from industrial uses is pollution. Pollution includes discharged solutes (chemical pollution) and discharged coolant water (thermal pollution). Industry requires pure water for many applications and utilizes a variety of purification techniques both in water supply and discharge.

Household

It is estimated that 8% of worldwide water use is for household purposes. These include drinking water, bathing, cooking, sanitation, and gardening. Basic household water requirements have been estimated by Peter Gleick at around 50 liters per person per day, excluding water for gardens. Drinking water is water that is of sufficiently high quality so that it can be consumed or used without risk of immediate or long term harm.

Such water is commonly called potable water. In most developed countries, the water supplied to households, commerce and industry is all of drinking water standard even though only a very small proportion is actually consumed or used in food preparation.

Recreation

Recreational water use is usually a very small but growing percentage of total water use. Recreational water use is mostly tied to reservoirs. If a reservoir is kept fuller than it would otherwise be for recreation, then the water retained could be categorized as recreational usage. Release of water from a few reservoirs is also timed to enhance whitewater boating, which also could be considered a recreational usage. Other examples are anglers, water skiers, nature enthusiasts and swimmers.

Recreational usage is usually non-consumptive. Golf courses are often targeted as using excessive amounts of water, especially in drier regions. It is, however, unclear whether recreational irrigation (which would include private gardens) has a noticeable effect on water resources. This is largely due to the unavailability of reliable data. Additionally, many golf courses utilize either primarily or exclusively treated effluent water, which has little impact on potable water availability.

Some governments, including the Californian Government, have labelled golf course usage as agricultural in order to deflect environmentalists' charges of wasting water. However, using the above figures as a basis, the actual statistical effect of this reassignment is close to zero. In Arizona, an organized lobby has been established in the form of the Golf Industry Association, a group focused on educating the public on how golf impacts the environment.

Recreational usage may reduce the availability of water for other users at specific times and places. For example, water retained in a reservoir to allow boating in the late summer is not available to farmers during the spring planting season. Water released for whitewater rafting may not be available for hydroelectric generation during the time of peak electrical demand.

Environmental

Explicit environmental water use is also a very small but growing percentage of total water use. Environmental water usage includes artificial wetlands, artificial lakes intended to create wildlife habitat, fish ladders , and water releases from reservoirs timed to help fish spawn, or to restore more natural flow regimes

Like recreational usage, environmental usage is non-consumptive but may reduce the availability of water for other users at specific times and places. For example, water release from a reservoir to help fish spawn may not be available to farms upstream.

CANAL BUILDING TECHNIQUE AND SCIENCE

Canals have to be designed and built to withstand enormous water pressure. In addition, they have to be watertight so that they will not leak and they have to be able offer protection from the risk of erosion caused by the flowing water.

The bottom of the canal is levelled off and compacted. A filter layer consisting of sand and gravel is then put in place and, on top of that, a sealing layer of fine soil, cement or asphalt. There is always a certain amount of leakage through the sealant and the filter ensures that the sealant does not get washed away. If fine soil is used as the sealing layer, it must be covered by an erosion resistant layer. It is highly important that the various layers are correctly compacted to avoid cracks in the sealing layer.

Even before the Revolutionary War gave New impetus to American expansionism, the colonial political and economic elites were deeply interested in the improvement of inland transportation. Vessels that plied offshore waters, small boats and rafts on the streams down to tidewater, and local roads and turnpikes served the immediate commercial needs of farmers and townspeople in the Atlantic coastal area. But the loftier dreams of planters, merchants, and political leaders—as well as of the common farmers who constituted by far most of the free population in British America—looked beyond the "fall line" that separated the rivers flowing to the coast from those that ran to the Ohio-Mississippi basin. A vast area for settlement and productivity—and riches—lay in the interior, and by the early 1790s demands for diffusion of New transport technologies and for investment in internal improvements were voiced frequently in both state and national political forums.

It was widely recognized that unless bulk agricultural commodities, which were the staples of a commercialized and expanding farm economy, could be carried cheaply. And over long distances, settlement and economic growth would be badly hampered in the region beyond the Appalachian Mountains. Then, too, there were opportunities for construction of short lines on the Atlantic seaboard to link already developed areas (coal mines, farming and lumber regions, and rising industrial sites), with the promise of immediate traffic and revenues. The latter were "exploitative" projects, tapping existing trade routes and resources; but the major east-west projects were "developmental," promoted with the goal of opening Newly or sparsely settled areas to economic opportunities. There was also a nationalistic or patriotic goal of canal promotion: to bind together far-flung sections of the young nation and to prove the efficacy of republican government.

And yet total canal construction in the United States up to 1816 totaled only 100 miles—the longest canal project being the Middlesex, which linked Boston's harbor with the farm region to the north. Other lines of some importance linked Norfolk, Virginia, to Albemarle Sound and connected the Santee River area to Charleston, South Carolina. Although many other canal projects were proposed up and down the Atlantic coast, progress was difficult because of shortages of capital and skepticism with regard to engineering feasibility projects. Moreover, regional or local jealousies notoriously worked against successful mobilization of governmental support in both the U.S. Congress and the state legislatures.

In the period from the mid-1820s to the Civil War, however, the United States underwent a vast expansion of canal construction, becoming the world's leading nation in both mileage of canals and the volume of tonnage carried on them. The canal lines were of crucial importance in the integration of a national economy, and they played a key role in the so-called "Transportation Revolution" that expedited both westward expansion and a robust industrialization process in the North and West.

Advantages, Disadvantages, and Construction Challenges

Canal technology proved especially attractive for several reasons. Since the 1760s, successful large-scale canal projects had been built in both Great Britain and France, and these canals had brought enormous economic advantages to the regions they served. The engineering advances pioneered in Europe gave American promoters confidence that they could build canals with equal success.

There was a downside to canal technology, too, though it was not always fully recognized in America. Difficult topography or uncertain water supply meant complex and highly expensive construction design. Canal building before the 1850s was mainly done with hand tools, augmented only by some primitive animal-powered machinery. A canal line had to be furnished with locks, permitting boats to pass through from one water level to another. The segments of line between the locks were of very gradual grade to permit controlled flow. At each lock, a gate at its higher level would be opened while the gate on the lower end was kept closed; once a boat entered the lock's chamber from the higher level, the upper gate was closed (holding the water flow back) while the lower gate was opened. As water ran out, the boat was carried down to the lower level, then passed through the open gate. For "upstream" movement, from the lower level to the higher, the process was reversed. The lock would be drained to the lower level, the boat would enter through the bottom gate, which was then closed, and water would then be admitted from the upper gate, lifting the boat up to the higher level. In steep areas, "flights" of locks, closely spaced, were necessary and often involved complex engineering; for transit of the boats, these series of locks meant a slow stretch and usually long waiting periods.

Locks varied in size. Lifts ranged from two to thirty feet, and there were great differences in the distances between gates as well as in the construction materials used. Masonry locks and metal or metal-trimmed gates were far more expensive—and more durable—than wooden gates and timber-supported rubble for the walls as found on some of the lines. The total rise and fall over an entire line was measured as "lockage," and served as an index of the difficulty of construction. For example, New York's Erie Canal route measured lockage of 655 feet, by contrast with Pennsylvania's lockage of 3,358 feet between Philadelphia and the Ohio River.

The size of the boats that could be accommodated, as well as the volume of water needed, were functions of the dimensions of the canal bed, or its "prism," as well as of the size of lock chambers. Prisms on American canals varied greatly, most of them ranging from forty to sixty feet in width at the top, with sloping sidewalls leading to a bottom of twenty-five to forty-five feet across. The Pennsylvania system was the most complex in engineering using inclined planes and steam-powered winches to drag boats out of the water and over some of the steepest hills.

To supply the line with flowing water, engineering plans had to include river connections, dams and reservoirs with feeder lines to the canals, and often massive culverts and aqueducts. Building the sidewalls to minimize loss of water through seepage was another challenging and expensive aspect of design. On many of the larger canals, such as the Ohio lines, engineers took advantage of fast-flowing feeder streams to design water-mill sites into the line.

Once a canal was in operation, moreover, maintaining navigation was a continuous challenge. Winter ice, droughts, floods, and breaches in the water-supply system would frequently cause navigational closings. Even in the best of circumstances, it was difficult to maintain regular schedules on the lines because of traffic bottlenecks at the locks and the continuous maintenance needed to keep water flowing.

Although steam-powered propeller craft were used on a few canal lines, this form of transport placed dangerous pressure on the canal walls. Hence, the use of horses or mules to haul canal boats was nearly universal, with the animals walking along the "towpath" alongside the line. Freight boats typically of 50-to 125-ton capacity operated at speeds of one to three miles per hour. On most lines they were owned by individuals or private companies, the line being a common carrier under the law.

In the short run, all the disadvantages of canal technology were more than offset by the cost savings for long-distance hauling, especially of bulk goods and produce. In the long run, however, innovations in steam technology and railroad engineering were destined to render many canals the losers in a New competitive age in transport that took shape in the late 1840s and the 1850s.

The Erie Canal

The great breakthrough came in 1817 with New York State's commitment to building the Erie Canal to connect the Hudson River at Albany with Buffalo on Lake Erie, a

Project far greater than any previously attempted in America. The Erie was important to subsequent canal development in several ways, most notably because it provided a model of public enterprise through its financing, administration, and implementation. The state raised capital through bond issues both in New York and in Europe, and supplemented these funds with tax revenues. Actual construction was overseen by a board of commissioners, some of whom were personally involved in the fieldwork, but the project became a celebrated "school for engineers," with most of the junior personnel learning their skills on the job under the tutelage of Benjamin Wright and James Geddes, two of less than twenty men who then constituted the profession in America. Many of the Erie engineers went on to direct canal surveys in other states.

The canal was divided into sections for purpose of construction, with private contractors taking on the work under the state engineers' supervision—a scheme that was emulated by nearly all the major canals subsequently built. It was an immediate commercial success once opened to its full length in 1825, leading the New York legislature to authorize a series of additional canals as well as the improvement and enlargement of the original line.

No nonmilitary enterprise in the United States had ever involved such expenditures as did the Erie, whose initial construction cost $6 million. The number of laborers employed was also unprecedented in any economic enterprise of the day. The state's construction expenditures energized local economies, giving part-time employment to farmers and creating sudden demand for stone, timber, mules, and oxen, and provisions for workers. Like canals and other public works throughout the country, moreover, the Erie attracted immigrant workers (mainly Irish and German) who were employed to do much of the most dangerous work.

The Erie's commercial impact on the rural countryside and on New York City's role as a center for trade with the interior and for exports to Europe—together with the rich stream of revenues from the tolls—heightened expectations everywhere in the country that other canals could produce equally spectacular fiscal and developmental results.

The Post-1825 Boom in Canal Building

Emulation of New York followed quickly. In 1825 Pennsylvania authorized a $10 million project, combining canal technology with the use of inclined planes. It was completed in 1834, tapping the Ohio Valley's farm country at Pittsburgh and giving Philadelphia trade advantages similar to those that its rival New York City had obtained from the Erie. The first of the western states to build a major line was Ohio, which authorized construction in 1825. Although still small in population and financial resources, Ohio, too, resorted to creation of a state enterprise and borrowed heavily both in the East and in Europe. Erie Canal engineers were brought in at first, but Alfred Kelley of Cleveland and Micajah Williams of Cincinnati,

Local entrepreneurs with no prior engineering experience, took principal charge of overseeing construction once the technical plans were adopted. Although administrative incompetence and corruption plagued the Pennsylvania project, Ohio's record was widely admired for its efficiency and strength of design. One line, the Ohio Canal, was completed in 1834 and extended from Cleveland on Lake Erie to Portsmouth on the Ohio River—the first water link between the Great Lakes and the great Mississippi-Ohio basin. A second line, completed in the mid-1840s, linked Cincinnati with Toledo to the north.

Other important lines begun or fully built prior to 1840 included the Delaware and Hudson Canal, a successful private line in the Pennsylvania coal country; the Delaware and Raritan Canal, also private, linking Philadelphia and New York; and the Chesapeake and Delaware Ohio Canal, which with substantial state support built a line, surveyed by the engineer William Strickland, through Maryland to link Baltimore with the Philadelphia port.

In the period 1815–1834, $60 million was invested in 2,088 miles of canals, with 70 percent of the funds coming from governmental sources, mainly the states. Most of the funds were borrowed at home and abroad. Also, Congress authorized the Army Engineers to conduct surveys for the states and federal companies; made some direct federal investments; and gave several million acres of public lands to Ohio, Indiana, and Illinois to subsidize their canal projects during this period.

In the decade following, 1834–1844, the "canal enthusiasm" continued to animate state governments and private promoters. Rivalries among states and competition among cities were intense, feeding the spirit of optimism. A new wave of canal construction followed, with the projects again heavily financed by loans from Europe and the eastern cities. Almost 1,300 miles of canal were built during this ten-year period. They cost $72 million, of which 79 percent represented public funds. In addition to major New state canal systems begun in Illinois and in Indiana (where the Wabash and Erie line would open another link for direct trade between the Ohio River and Lake Erie), and in Illinois, three of the pioneering state projects—the Erie, Ohio's two main canals, and the Pennsylvania system—were further expanded to satisfy sections of their states that had been left out of the original system designs. As the New canals were generally of larger dimensions than the first ones to be built, the carrying capacity for canal traffic doubled between 1834 and 1844. Until 1839, conditions of prosperity and expansion sustained the canal-building movement, and expenditures for the New canals stimulated overall economic growth.

PREHISTORIC ENGINEERING

Nobody in the New World moved water more efficiently than the Hohokam Indians. The canals they built to sustain their large and complex agricultural society are still considered an engineering marvel.

The Hohokam people lived around the Mesa area of Central Arizona for almost 1,500 years. They appeared around 1 CE and disappeared about 1450. Their agricultural society flourished around the middle Gila River and Salt River. These sources allowed them to farm beans, squash, corn, and cotton. The Hohokam were skilled agrarians, managing the soil to replace lost nutrients. Additional food came from extensive gathering of wild plants, dry farming agave, and hunting deer and small animals.

Around 600 CE the Hohokam people began to construct irrigation canals. They excavated trenches up to 12 feet deep by hand, using a digging stick. The trenches drew off Salt River water and fanned into a network of smaller canals that brought a steady supply of water to fields. Between 1100 CE and 1450 CE, 500 miles of canals irrigated 110,000 acres. The food produced by this advanced irrigation system is believed to have supported up to 80,000 people—the highest population density in the prehistoric Southwest.

Hohokam canals possessed remarkable features that still stand as examples of engineering genius. First, they were wide at the mouth and carefully tapered, getting smaller as secondary branches drew water from the main channel. By shrinking the channel size as the flow decreases, the Hohokam were able to stabilize flow rate. A steady flow rate is key to creating a functional irrigation system. Water that moves too fast carries sand and silt that will eventually block the canal. Water that moves too slowly will not reach its destination. Second, canals covered miles of territory on a relatively steady grade, avoiding hills and valleys that could affect flow rate.

Though the Hohokam people are most celebrated for their canals, they made fire-hardened pottery and figures and jewelry from shell and turquoise. They built increasingly elaborate structures, such as the Big House in Casa Grande Ruins National Monument. The society also constructed large earthen platform mounds of unconfirmed purpose. Located in the center of villages, the mounds may have been used for housing elite members of the society or for ceremonial activities. Hohokam rock art, in the form of petroglyphs and pictographs, is found in mountain ranges and rocky areas throughout the region.

Mysterious Migration

The Hohokam disappeared from their villages near the Salt and Gila Rivers in 1450 CE. Their descendants survive as members of the Tohono O'odham Nation.

The three centuries leading up to the Hohokam's disappearance around 1450 CE had been a period of migration and social change, with old settlements and prestigious structures constructed in the Salt and Gila Basins being abandoned. The reasons for this mass abandonment remain unknown to this day. Perhaps disease or war threatened the population. Perhaps flood damaged the canal infrastructure, or drought brought hardship. There is also speculation that a great leader died, causing the people to drift away and migrate. We may never discover why the Hohokam decided to leave the area they had lived in for almost 1,500 years. But their descendants live on as members of the Tohono O'odham Nation.

Reportedly, canals were built in Egypt as much as 6,000 years ago. The first canal designed to accommodate seaworthy ships was built by the Pharaoh Sesosteris I4,000 years ago, linking the Red Sea to the Nile River and thus to the Mediterranean Sea. Remnants of this canal, which was alternately neglected and rebuilt half a dozen times, suggest that it was about 150 ft (46 m) wide, 16 ft (5 m) deep, and 60 mi (97 km) long.

The oldest canal still in use was begun in China in the fourth century B.C. for the purposes of collecting grain taxes and transporting troops. Rebuilt and enlarged about A.D. 600, the channel became known as the Grand Canal; its latest extension, completed about A.D. 1280, brought its length to 1,114 mi (1,795 km). It remains the longest navigable canal in the world.

For centuries, canals were dug using only hand tools. While building the Canal du Midi between 1665 and 1681 to link the Atlantic

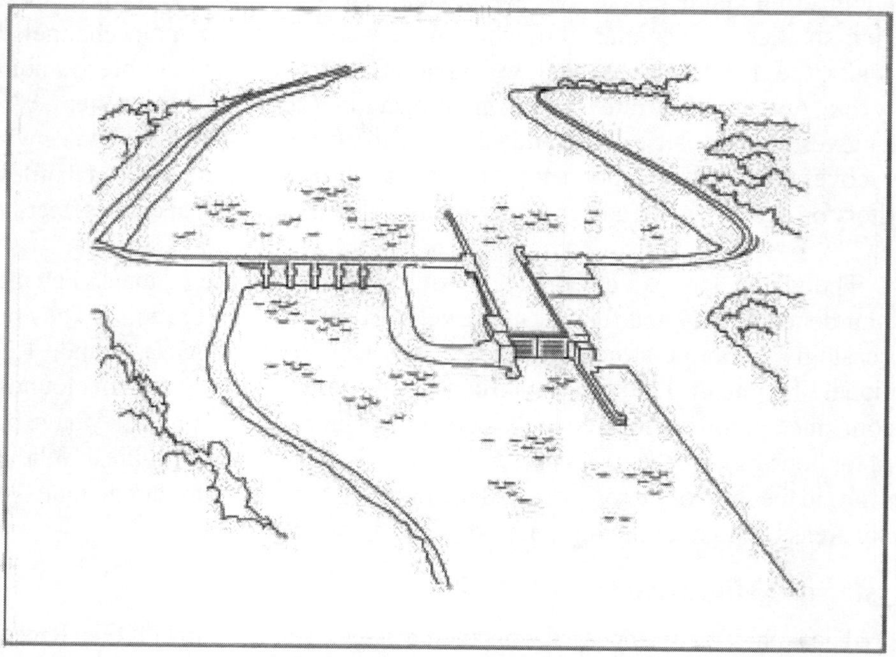

An aerial view of a canal and lock system.

Ocean and the Mediterranean Sea across southern France, gunpowder was first used to blast a tunnel. The invention of steam-powered machinery at the end of the seventeenth century helped fuel the "canal eras" in Europe and North America in the eighteenth and nineteenth centuries during which thousands of miles of canals were built. Horse-drawn plows continued to be used especially for final trimming of the channel to desired slopes and depths. Full mechanization of canal construction was achieved in 1946, when an American company built the first canal trimming and lining machines.

Locks

Flash locks, the first attempts to carry boats over difficult elevation changes on rivers or canals, date from the third century B.C. Greek historian Diodorus Siculus, writing 200 years later, described how Ptolemy II had improved the Nile-to-Red Sea canal by building a type of lock. Consisting of a single gate, flash locks carried boats downstream on a rush of water; boats headed upstream could be pulled forward while floating on the torrent released by opening the gate.

Flash locks were hazardous, and they used large amounts of water. A significant improvement came in A.D. 984, when the first double-gate lock was built on China's Grand Canal. Also called a pound lock (because it impounds water) this was the predecessor of modern conventional locks. Its gates were panels that lifted vertically. A similar lock with vertical-lift gates built in the Netherlands in 1373 also controlled its water level by partially opening either the upstream or the downstream gate. Pound lock operation was greatly improved in 1485 when an Italian lock was built with smaller, valve-controlled openings in the gates.

Leonardo da Vinci invented the miter gate in 1480. Two gates, each more than half as wide as the lock, swing on vertical hinges. In the open position, they are flush with the lock's walls. In the closed position, they meet in a V pointed upstream so that the higher water level presses against them to promote a tight seal.

The base of the upstream gate in a lock is higher than the base of the downstream gate. If the lock was very deep, water filling the chamber through an opening in the upper gate would create turbulence or even swamp a boat in the chamber. This problem was solved in France during the seventeenth century, when valve-controlled water channels for filling and emptying a lock were built

A lock is a rectangular chamber with gates at both ends; with both gates closed, the water level within the lock can be adjusted to match the canal water level on either side. Thus, a vessel entering the lock can be raised or lowered in order to enter the next level canal section.

Until the early nineteenth century, lock gates were made of wood, and lock chambers were lined with wood, stone, bricks, or turf. In 1827, cast iron was first used to build both a lock and its gates in Cheshire, England. Following the development of the Bessemer process for mass production of steel, that material was utilized for construction of locks and gates, both as primary elements and as reinforcing bars for concrete.

RAW MATERIALS

Waterproof linings keep a canal's water from seeping into the ground. For many years, the best choice was puddle, a mixture of sand, clay, and water that dried to a waterproof state. Modern materials and additives that are more durable include concrete, fly ash, bentonite, bituminous materials, and plastic sheeting.

Locks are usually made of concrete, occasionally lined with steel. If construction of the lock exposes bedrock, the floor need not be lined. The gates are made by welding together steel plates and reinforcement beams. The vertical edges of the gates are fitted with effective sealing materials such as white oak. In 1999, a French company developed lock gates made of glass-fiber reinforced plastic laminate mounted in stainless steel frames.

Design

Early canals followed the most level surface route possible because large-scale earth-moving was so difficult and expensive. Better excavation equipment and lock construction capabilities permit the construction of shorter, more direct canal routes. Because of geographic obstacles, sections of some canals are built in tunnels or on aqueducts (water-carrying bridges).

It is important to fill and empty a lock chamber while producing minimal water turbulence. Modern designs place sluices in the gate sills or in the chamber's floor or walls. A lock may also be equipped with a submerged bubbler that releases air below the gate closure area; the resulting gentle turbulence keeps the area clear of debris that might prevent the gate from sealing properly.

Various gate designs are available for modern locks, and different types may be used on the upstream and downstream ends of a single lock. Miter gates are one of the most popular choices. Another common choice is the Tainter gate, a curved plate that rotates vertically. In this efficient design, which is used for valves in water-transfer culverts as well as main lock gates, water pressure actually assists in the gate's rotation. Flat gates that slide up, sideways, or down into the lock floor are other options, as are flat gates that are hinged at the bottom and curved gates that rotate horizontally into wall recesses.

William C. Gorgas was born October 3, 1854, near Mobile, Alabama. In 1875 Gorgas received a bachelor of arts degree from the University of the South. Desiring a military career he decided to enter the Army with a medical degree. After graduating from Bellevue Medical College in New York and interning at Bellevue Hospital, Gorgas was appointed to the Medical Corps of the U.S. Army in June 1880. Tours of duty followed in Texas and North Dakota, with nearly 10 years at Fort Barrancas in Florida—Gorgas was assigned this yellow fever area because he had previously had the disease and was immune.

After Havana, Cuba, was occupied by American troops in 1898, Gorgas took charge of a yellow fever camp at Siboney, soon becoming chief sanitary officer of Havana. Acting on information that a certain mosquito carried yellow fever, Gorgas quickly destroyed that mosquito's breeding ground, ridding the city of yellow fever.

In 1904, work commenced on the Panama Canal. Gorgas went to the Canal Zone to take charge of sanitation, succeeding in Panama and Colon. Gorgas came to be generally regarded as the world's foremost sanitary expert, and a number of foreign governments and international commissions sought his aid. His book Sanitation in Panama quickly became a classic in the public health field. In 1914 he was appointed surgeon general of the Army, serving in that capacity until his retirement four years later. Gorgas died on July 3, 1920, and is buried in Arlington National Cemetery.

The Manufacturing Process

Canal

- 1 A thorough survey of the canal route must be made, not only for correct alignment but also for accurate depths of cuts and fills.
- 2 Primary excavation is done with earth-moving equipment like bulldozers and excavators.
- 3 A crawler-track mounted trimming machine is used to remove the final 12-18 in (30-46 cm) of soil to create the desired wall slopes and flat bed for the canal. Excavated soil is carried by conveyor and loaded onto trucks that will haul it away.
- 4 A layer of highly permeable material may be spread over the walls and floor to promote drainage of groundwater under the canal.
- 5 Grids of reinforcing steel are built and l1ifted into place on the canal walls and floor. Concrete blocks hold the grids above the surface so concrete can flow under them.
- 6 Concrete is mixed, vibrated to remove air pockets, and applied with a slipform machine, encasing the reinforcing steel. Slipforming means pouring concrete between the canal surface and a form (mold) that slowly moves forward; the concrete hardens quickly enough to retain its shape by the time the form moves out of contact with it. Specialized machines place transverse and longitudinal expansion joints in advance of the slipform machine.

Lock

- 7 A temporary cofferdam is built around the proposed lock site. Steel sheet piles are driven into the ground to form a series of adjacent, vertical cells that extend above the waterline. The cells are filled with sand. Water is pumped out of the enclosed space to create a dry construction area. The lock site is excavated. If necessary, piles are driven into the ground to support the lock structure.
- 8 Wooden forms are built to shape the floor and walls of the lock. Space for culverts and valve chambers is included within the forms, as are slots for the gate hinges and recesses to contain open gates. Additional forms are built to shape the approach walls that will guide vessels into the lock.
- 9 After reinforcing steel cages are constructed in the forms, concrete is poured. When the concrete has cured (hardened), the forms are removed.
- 10 Control valves for filling and emptying the lock are installed, along with hydraulic and mechanical equipment for operating the valves and gates.
- 11 Gates are prefabricated and shipped to the site. Very large gates may be

shipped in sections, which are welded together as they are installed in the lock.

- 12 Accessories such as guardrails, mooring posts, and escape ladders are installed on the lock walls.

Byproducts/Waste

Disposing of the excavated material is one of the challenges of canal and lock construction. It may be used to construct embankments or be spread over the surrounding countryside where it is carefully landscaped for erosion control as well as appearance. Construction of the Divide Cut, a 29-mi (46-km) canal on the Tennessee-Tombigbee Waterway, in the early 1980s required the disposal of 150 million cubic yards (115 million cubic meters) of dirt.

During lock construction, excavated material can be used to fill cofferdam cells. When the cofferdam is removed, the material may be used to fill in behind the riverbank side of the lock wall.

Lock operation uses large amounts of water; for example, it takes 3.5 million gal (13 million L) of water to fill a 600-ft (180-m) long, 110-ft (34-m) wide lock with a 7-ft (2.1-nm) lift. On some canals and rivers, water supplies are limited and conservation is important. Rather than being released downstream as the lock is emptied, some of the water may be diverted to a side pond, where it can be stored and used to help fill the chamber for its next operation.

The Future

Because of the significant need to recondition or replace locks as well as build new facilities, the U.S. Army Corps of Engineers sponsors and conducts research on water-way construction. Scale models of proposed designs can be tested at its Waterways Experiment Station. Through its Innovations for Navigation Projects (INP) Research Program, it supports the development of new ways to build, repair, and operate canals and locks. Recent topics included improvements in underwater concrete and grout placement and development of low-density, high-strength concrete for building modular lock sections that can be installed without construction of a cofferdam. Other current research topics involve protection devices for lock walls and gates from barge impacts, improved designs for water intake and discharge systems, better equipment and control devices for operating gates, and techniques for enlarging (rather than replacing) existing locks that have inadequate capacity.

Canal - Construction and Operation

Engineers who are designing a canal must take several things into consideration. They must formulate the dimensions of the canal to accommodate the numbers and sizes of ships that are predicted to use the waterway. Natural obstacles in the path of the canal, such as rock formations, must also be modified, removed, or avoided. There must be adequate vertical clearance above the canal and the clearance afforded by pre-existing bridges must be able to accommodate the vessels that will use the canal. Finally, engineers must decide the scale and location of associated structures, such as bridges, tunnels, and locks.

The paths of most canals are affected by variations in the levels of terrain. Engineers compensate for these variations with either locks or inclined planes. A lock is a segment of the waterway that is closed off by gates at either end. When a vessel enters the lock, the front gate is already closed. The back gate is then closed behind the vessel, and the water level within the lock is raised or lowered to the level of the water on the outside of the front gate. Valves on the gates control the level of the water. Whereas locks are the most common means of compensating for elevation changes, the procedure is slow and uses a large amount of water. Inclined planes can be used to elevate and lower smaller vessels; they use no water and often allow more rapid passage of vessels. When vessels reach certain stations along the waterway, they are pulled out of the water and moved on trucks up or down the plane.

Canal operators must monitor the canal's supply of water. If the natural supply of water at the upper end of the canal is deficient, it must be supplemented by water pumped into the reservoirs. If nature supplies the reservoir with excess amounts of water, some of the water must be diverted from the canal. Otherwise, excess water may strengthen the current and disrupt canal operations.

Although canals are among the oldest civil works, they play a major role in commerce because they are by far the least expensive form of inland transportation yet devised.

Steps to complete Canal Construction

Identity cards had been issued to 2.02 lakh agriculture labourers under Tamil Nadu Agriculture Labourers Social Security scheme in the district so far, said the Collector, Har Sahay Meena.

Distributing identity cards to farm labourers at a function held at Theni-Allinagaram Cooperative Market here on Friday, he said agri-labourers and tiny and small farmers aged between 18 and 65 years had been included in the scheme.

Small farmers having 2.5 acres of wetland and five acres of dryland were eligible for this scheme.

The eligible beneficiaries need not pay any amount for joining this scheme. Each beneficiary, in case of death, would get Rs.1 lakh as compensation. A sum of Rs.50,000 would be given as compensation to the beneficiaries who had lost their limbs or eyes in accidents. The injured would get Rs.25,000 and family member of the beneficiary would receive Rs.10,000 in case of natural death of the beneficiary.

Marriage assistance of Rs.2,000 would be given to male member and Rs.5,000 to female member. Maternity assistance of Rs.2,500 would be given to women members and educational assistance ranging from Rs.1000 to Rs.6,500 to the wards of members.

All efforts would be made to complete the construction of 18th canal as early as possible, he said, adding that the work at Cumbum and Gudalur had been completed.

"We have been waiting for approval from the Forest Department for acquisition of land to construct a dam across the Vairavanar river."

Mr. Meena ordered the officials to expedite acquisition of land for constructing a canal at Melagudalur and Uthamapalayam areas.

On completion of the work, 44 old irrigation tanks and four new tanks would receive water and 4,704 acres would be irrigated.

This canal would pass through Gudalur, Cumbum, Uthamapuram Pudhupatti, Kombai and Thevaram, mostly dry areas, and connect with the Sutha Ganga stream.

DESIGN OF LINED CANALS

The lined canals are not designed making use of Lacey or Kennedy Theory because the section is rigid. Generally Manning's equation is used in design. To carry a certain discharge number of channel sections may be designed with different bed widths and side slopes. But it is clear that each section is not equally good for the purpose.

The section to be adopted should be economical and at the same time it should be functionally efficient. It has been found that the most suitable cross-section of a lined canal is a circular section with sloping sides. That is, the bed is not flat but it is an arc of a circle. This arc is tangential to the sloping sides.

Side Slopes

The side slope is selected in such a way that it nearly equals the angle of repose of the soil in the subgrade. Care is taken to ensure that no earth pressure is exerted on the back of the lining. From the knowledge of hydraulics it is clear that the section is economical when cross- sectional area is maximum for minimum wetted perimeter.

This condition is achieved when the centre of an arc lies at FSL of the canal. This section is also efficient in the sense that as the velocity of flow is higher silt carrying is also higher than a wide and shallow section. Thus the problem of silting is completely eliminated and functioning is efficient.

It may be mentioned here that such section with circular bed may be designed up to a discharge of 85 m/sec. When the discharge is more than 85 m/sec the section best suited is one with a flat bed and sloping sides with rounded corners. This section is certainly better than trapezoidal section, because it is more stable and economical to construct. Dimensions of a canal section with circular bed may be obtained from the equations given below. Figure 10.2 shows the canal sections of this type with two standard side slopes.

When $r = 3.6$ m or less, side slopes may be taken 1: 1

When $r = 3.6$ m side slopes may be taken 1.25: 1

Velocity of Flow:

Mean velocity of flow may be calculated using Manning's formula.

V = 1/N .R. S

where N is coefficient of rugosity and may be taken as 0.018

S is the slope of bed and expressed in fall of bed in m in 10,000 m length

For the given value of N formula may be reduced to

V = 0.556 R. S

where V is velocity of flow in m/sec;

R is hydraulic mean radius in m; and

S is bed slope expressed in metres/metre length,

since (10,000) is merged in the constant 0.556.

Following Table 10.1 gives the sectional data for the two side slopes:

Dimensions of a canal section with flat bottom may be obtained from the equations given below. Figure 10.3 shows the cross-sections of this type with two standard side slopes. It may be mentioned here that generally maximum velocity of flow is recommended to be 2 m/sec and Lacey's 'f is limited to 1.2. When r exceeds 55 m or when discharge is more than 85 m/sec, the canal should be designed with a flat bottom with fillets at corners. The Indian Standard 4745 in fact does not suggest lined canal with circular bottom.

Velocity of Flow

Mean velocity of flow may be calculated from Manning's formula for value of N = 0.018.

Then V = 0.556 R. S

Following Table 10.2 gives sectional data for the two side slopes:

It may be remembered that lined canal could be given steep slope to achieve recommended velocity of 2 m/sec and value of 'f' about 1.2. Critical velocity ratio is not applicable to lined canals. But to avoid possibility of sitting CVR should be aimed at more than unity.

Coefficient of Rugosity (N)

In general practice for lined canal average value of N may be taken as 0.018. For different types of linings the value of W varies. The values given for straight channels in Indian Standard 4745 are given in Table 10.3. When the alignment is not straight loss of head increases and a small increase in the value of W may be made to allow for additional loss of energy.

Freeboard

Freeboard is measured from the (FSL) full supply level to the top of lining. For lined canals having less than 10 cumec discharge 0.6 m free board is recommended. For bigger lined canals freeboard not less than 0.75 m is generally provided.

Bank Widths

The Indian Standard recommended following values for bank widths for main and branch canals:

Main canal in cutting and filling = 8.0 m
Branch canal in cutting = 6.5 m
Branch canal in filling, left bank = 6.5 m
Right bank = 5.0 m

Suitable provision of dowel, roadway, catch water drain, under drainage has to be made for every lined canal. Figure 10.4 shows three typical cross-sections of the lined canal in which provision of various components is illustrated.

The canal sections shown have trapezoidal shape. For circular bottomed canal arrangement will be similar. It may be noted that angle 0 shown in the section varies with the side slopes adopted. 9 for side slope 1: 1 has 45° value and for 1.25: 1 is 38° 402 as shown in Fig. 10.3. Maximum height of the spoil bank is limited to 6 m when the excavation is done by machinery. In case work is accomplished manually maximum height is to be restricted to 4 m only. Also when a canal section in filling involves more than 10.5 m filling each section shall be designed and tested for stability.

6

Dams

An arch-gravity dam, curved-gravity dam or arched dam is a dam with the characteristics of both an arch dam and a gravity dam. It is a dam that curves upstream in a narrowing curve that directs most of the water against the canyon rock walls, providing the force to compress the dam. It combines the strengths of two common dam forms and is considered a compromise between the two. A typical example is the Hoover Dam. A gravity dam requires a large volume of internal fill. An arch-gravity dam can be thinner than the pure gravity dam and requires less internal fill.

Arch-gravity dams are massive dams of reinforced concrete that resist the thrust of water by their weight pushing down using the force of gravity. A gravity dam is constructed so that the dam's massive weight resists the pressure of the water against it.

At the same time an arch-gravity dam incorporates the arch's curved design that is so effective in deflecting the water in narrow, rocky locations where the gorge's side are of hard rock and the water is forced into a narrow channel. Therefore the span needed for the dam is narrow; the dam's curved design effectively holds back the water in the reservoir using a lesser amount of construction material.

When properly situated on an appropriate site, the arch-gravity dam, combining the load resisting qualities of both a gravity dam and an arch dam, inspires the most confidence in the public at large because of its appearance of massive permanence. However, curving a gravity dam may make it look stronger, but some of this effect may be psychological factors in operation in making a choice of dam style.

Gravity dam

The gravity dam is a massive structure made of masonry or concrete with earth and rock fill. Gravity dams use their own weight to resist opposing forces and as such require a hard bedrock foundation. They rely on their great weight and size for stability. The gravity dam is the most commonly built dam in the world. One of the reasons for this is the simplicity of design. Unfortunately it requires great quantities of material to construct.

Arch dam

An arch dam is a thin, curved concrete or masonry dam structure, with the concave side of the curve downstream. Therefore, the force of the water against the canyon wall squeezes the arch, compressing and strengthening its structure and pushing it into the ground. An arch dam is a good dam type for a narrow gorge in a mountainous area with steep walls of rock. It generally has steel rods or prestressed steel cable reinforcements and therefore requires less concrete than a gravity or arch-gravity dam. However, the bedrock in the foundation and abutments must be sound to withstand the pressure of the water.

Arch dams with more than one contiguous arch or plane are described as multiple arch dams. A double arch dam has two contiguous arches. A dam that is curved in both its horizontal and vertical planes may be called a dome dam.

Masonry dams are dams made out of masonry; mainly stone and brick. They are either the gravity or the arch type. The largest masonry dam of the world is Nagarjunasagar Dam in India.

Nagarjuna Sagar Dam

Nagarjuna Sagar Dam is the world's largest masonry dam built across Krishna River in Nagarjuna Sagar, Nalgonda District of Andhra Pradesh, India. It is downstream to the Nagarjuna Sagar reservoir with a capacity of up to 11,472 million cubic metres which is the world's largest man-made lake with a concrete wall of that measures 6 ft (1.8 m). thick. The dam is 490 ft (150 m). tall and 16 km long with 26 gates which are 42 ft (13 m). wide and 45 ft (14 m). tall.

It is one of the earliest irrigation and hydro-electric projects in India. The dam provides irrigation water to the Nalgonda District, Prakasam District, Khammam District and Guntur District.

The proposal to construct a dam to use the excess waters of the Krishna river was put forward by the British rulers in 1903. Siddeswaram, Hyderabad and Pulichintala were identified as the suitable locations for the reservoirs. The perseverance of the Raja of Muktyala paved way for the site identification, design and construction of the dam.

The dam water was released by the then Prime Minister's daughter, Indira Gandhi in 1967. The construction of the dam submerged an ancient Buddhist settlement, Nagarjunakonda, which was the capital of the Ikshvaku dynasty in the 1st and 2nd centuries, the successors of the Satavahanas in the Eastern Deccan.

Excavations here had yielded 30 Buddhist monasteries, as well as art works and inscriptions of great historical importance. In advance of the reservoir's flooding, monuments were dug up and relocated. Some were moved to Nagarjuna's Hill, now an island in the middle of the reservoir. Others were moved to the mainland.

EFFECT OF THE PROJECT

The project benefited farmers in the districts of Guntur, Prakasam, Krishna, Nalgonda and Khammam. The right canal (a.k.a Jawahar canal) is 203 km long and irrigates 1.113 million acres (4,500 km^2) of land. The left canal (a.k.a Lalbahadur Shastri canal) is 295 km long and irrigates 0.32 million acres (800 km^2) of land in Nalgonda and Khammam districts of Telangana region. The project transformed the economy of above districts. 52 villages were submersed in water and 24000 people were affected. The relocation of the people was completed by 2007.

Power Generation

The hydroelectric plant has a power generation capacity of 815.6 MW with 8 units (1x110 MW+7x100.8 MW). First unit was commissioned on 7 March 1978 and 8th unit on 24 December 1985. The right canal plant has a power generation capacity of 90 MW with 3 units of 30 MW each. The left canal plant has a power generation capacity of 60 MW with 2 units of 30 MW each.

The dam is constructed on the border of Guntur and Nalgonda districts. The dam also provides drinking water to the Nalgonda town.

Statistics

- Catchment Area : 215000 km^2 (83012 sq mi)
- Masonry dam
  - Spillway of dam : 471 m
  - Non-over flow dam : 979 m
  - Length of Masonry dam : 1450 m
  - Maximum height : 125 m
- Earth dam
  - Total Length of Earth dam : 3414 m
  - Maximum height : 28 m

Arch dam

An arch dam is a type of dam that is curved and commonly built with concrete. The arch dam is a structure that is designed to curve upstream so that the force of the water against it, known as hydrostatic pressure, presses against the arch, compressing and strengthening the structure as it pushes into its foundation or abutments. An arch dam is most suitable for narrow gorges or canyons with steep walls of stable rock to support the structure and stresses.

Since they are thinner than any other dam type, they require much less construction material, making them economical and practical in remote areas. The development of arch dams throughout history began with the Romans in the 1st century BC and after several designs and techniques were developed, relative uniformity was achieved in the 20th century.

The first known arch dam, the Glanum Dam, also known as the Vallon de Baume Dam, was built by the Romans in France and it dates back to the 1st century BC. The dam was about 12m high and 18m in length. Its radius was about 14m and it consisted of two masonry walls. The Romans built it to supply nearby Glanum with water.

The Monte Novo Dam in Portugal was another early arch dam built by the Romans in 300 AD. It was 5.7m high and 52m long with a radius of 19m. The curved ends of the dam met with two winged walls that were later supported by two buttresses. The dam also contained two water outlets to drive mills downstream.

The Dara Dam was another arch dam built by the Romans in which the historian Procopius would write of its design: "This barrier was not built in a straight line, but was bent into the shape of a crescent, so that the curve, by lying against the current of the river, might be able to offer still more resistance to the force of the stream."

The Mongolians also built arch dams in modern-day Iran. Their earliest was the Kebar Dam built around 1300 which was 26m high, 55m long and had a radius of 35m. Their second dam was built around 1350 and is called the Kurit Dam. After 4m was added to the dam in 1850, it became 64m tall and remained the tallest dam in the world until the early 20th century. The Kurit Dam was of masonry design and built in a very narrow canyon. The canyon was so narrow that its crest length is only 44% of its height. The dam is still erect even though part of its lower downstream face fell off.

The Elche Dam in Elche, Spain was a post-medieval arch dam built in the 1630s by Joanes del Temple and the first in Europe since the Romans. The dam was 26m high, 75m long and had a radius of 62m. This arch dam also rest on winged walls that served as abutments.

In the 20th century, the world's first variable radius arch dam was built on the Salmon Creek near Juneau, Alaska. The Salmon Creek Dam's upstream face bulged upstream which relieved pressure on the stronger curved lower arches near the abutments.

The dam also had a larger toe which off-set pressure on the upstream heel of the dam which now curved more downstream. The technology and economical benefits of the Salmon Creek Dam allowed for larger and taller dam designs. The dam was therefore revolutionary and similar designs were soon adopted around the world particularly by the U.S. Bureau of Reclamation.

Arch dam designs would continue to test new limits and designs such as the double and multiple-curve. The Swiss engineer Alfred Stucky and the U.S. Bureau of Reclamation would develop a method of weight and stress distribution in the 1960s and arch dam construction in the United States would see its last surge then with dams like the 143m double-curved Morrow Point Dam in Colorado. By the late 20th century, arch dam design reached a relative uniformity in design around the world.

Currently, the tallest and largest arch dam in the world is the 272 m Inguri Dam in Georgia which was completed in 1980. The longest multiple arch with buttress dam in the world is the Daniel-Johnson Dam in Québec, Canada. It is 214 m high and 1,314 meters long across its crest. It was completed in 1968 and put in service in 1970. China is currently constructing the Jinping 1 Dam in Sichuan which when complete will be world's tallest arch dam at 305m.

Design

Arch dams require a high level of stress and force analysis in order to create a sufficient design. The main force against an arch dam is the hydrostatic pressure provided by the reservoir behind it, uplift which is water pressure beneath the dam, the weight of the dam itself and finally all the forces combined. Other forces that affect a dam include, but are not limited to, temperature, chemical reactions, settling, silt accumulation and earthquakes.

Most often, the arch dam is made of concrete and placed in a "V" shaped valley. The foundation or abutments for an arch dam must be very stable and proportionate to the concrete. There are two basic designs for an arch dam: constant radius which have constant radius of curvature and variable radius dams which have both upstream and downstream curves that systematically decrease in radius below the crest.

A dam that is double-curved in both its horizontal and vertical planes may be called a dome dam. Arch dams with more than one contiguous arch or plane are described as multiple-arch dams. Early examples include the Roman Esparragalejo Dam and Muro Dam with later examples such as the Daniel-Johnson Dam (1968) and Itaipu Dam (1982). At an increased cost, an arch dam can have steel rods or pre-stressed steel cable reinforcements and therefore require less concrete than does a gravity dam or arch-gravity dam. As arch dams are commonly constructed in remote mountainous areas, less material is beneficial when it has to be delivered with difficulty to the site. However, as a result of the failure of the Gleno Dam shortly after it was constructed in 1923, the construction of new multiple arch dams has become less popular.

Arch dams are generally classified as thin, medium or thick. If width of an arch dam's base is 2/10 of the dam's height or less it is thin. Any base and height ratio greater than 2/10 but less than 3/10 is considered medium. Thick is a ratio of 3/10 or greater. Additionally, an arch dam's base is historically just as thick as the crest but often it is twice as thick as the crest. Contraction joints are normally placed every 20 m in the arch dam and are later filled with grout after the control cools and cures.

Examples of arch dams

- Inguri Dam
- Buchanan Dam (example of Multiple-Arch type)
- Glen Canyon Dam

- Hoover Dam
- Idukki Dam
- Karun-3 dam
- Daniel-Johnson Dam

Daniel-Johnson Dam

The Daniel-Johnson Dam, formerly known as Manic-5, is a multiple arch buttress dam on the Manicouagan River which creates Manicouagan Reservoir. The dam is composed of 14 buttresses and 13 arches and is 214 km (133 mi) north of Baie-Comeau in Quebec, Canada. The dam was constructed between 1959 and 1970 for the purpose of hydroelectric power production and supplies water to the Manic-5 and Manic-5-PA power houses. The dam is 214 m (702 ft) tall, 1,314 m (4,311 ft) long and contains 2,200,000 m3 (2,880,000 cu yd) of concrete, making it the largest dam of its type in the world.

The dam was named after Daniel Johnson, Sr., the 20th Premier of Quebec who was responsible for starting the project. Johnson died on 26 September 1968, on the day he was to preside over the scheduled inauguration of the dam. The facility is owned and operated by Hydro-Québec.

During the summers of 1919 and 1920, hydrological studies were conducted on the Manicouagan and Outardes rivers, flowing into the St. Lawrence near the town of Baie-Comeau. The combined flow was then estimated at 40 million cubic meters, making it one of the largest hydrological systems in Canada. Although harnessing of this potential was considered interesting, its distance from major load centers and the lack of roads in the area were identified as major drawbacks. In addition, building dams in the wilderness was considered too costly.

After the Second World War, the discovery of large iron deposits on the North Shore and the increased forestry activity led to a rapid development of the region. The area's largest cities, Sept-Iles and Baie-Comeau, were now linked to the rest of the province by a road. At the same time, industrial development in southern Quebec required a larger electric supply. Improvements in long distance electric transmission technologies were also a consideration at the time. Significant advances in the field, including the construction of two 315-kilovolt lines, between the Bersimis complex, west of the Manicouagan system, and Montreal, (completed in 1956), lifted another obstacle.

In 1955, Hydro-Québec launched an 5-year extensive assessment of the Manicouagan's suitability. These studies demonstrated the exceptional potential of the river and stressed the advantages of building a multi-dam system in order take full advantage of the terrain and water flows. The data collected at the time was so promising that Hydro-Québec did not wait for the working group to submit their final recommendations.

By 1959, a decision was made and construction of a 210 kilometres (130 mi) access road from Baie-Comeau was started. The original project, the Manicouagan-

Outardes project, included the construction of five dams on the Manicouagan River (Manic-1, Manic-2, Manic-3, Manic-4 and Manic-5) and three on the Outardes River (Outardes-2, Outardes-3 and Outardes-4). However, a miscalculation prevented the construction of the Manic-4 dam and powerhouse because engineers realized early on that it would encroach on the Manic-3 reservoir.

In 1959, the Premier of Quebec, Maurice Duplessis, wanted a U.S. firm to construct the dam but Quebec's Minister of Water Resources at the time, Daniel Johnson was against the idea. Johnson believed that the skill of a Canadian firm and workers was well suited for the construction of such a complex dam. Duplessis' successor in 1959, Paul Sauvé agreed with Johnson and he continued to pave the way for the dam. Eventually, engineers would choose André Coyne's design for a multiple arch buttress dam as the most suitable and economical.

Construction

Construction on the dam's support facilities began in 1959 and the river diversion tunnels and foundation preparation commenced in 1960. To divert the Manicoagan River, workers blasted and dug two 2,000 ft (610 m) long and 45 ft (14 m) diameter tunnels through the gorge's solid granite west wall.

Workers used "jumbo" drilling platforms and the tunnels progressed 14 ft (4.3 m) each shift. To facilitate the river's diversion, two cofferdams were built. The first was a concrete arch upstream from the dam site which would block the river, forcing it into the diversion tunnels. The second cofferdam was downstream and prevented water from flowing back into the construction site. Each cofferdam was built on alluvial deposits (loose soil) so they needed to be watertight which was accomplished with either a grout curtain or deep supporting piles.

Once the river was diverted, workers pumped the remaining water out between the two cofferdams in order to prepare the site for construction. Once drained, workers excavated the alluvial deposits between the cofferdams creating a long pit that was 150 ft (46 m) deep at the center. This pit was then filled with concrete in the summer of 1962. To prevent seepage in the dam's foundation, a grout curtain was injected in the bedrock and a drainage network with 2,400 ft (730 m) of tunnels was constructed just downstream of the dam to collect water that may seep through.

After the extensive preparatory works were complete, the first concrete for the dam was poured on October 3, 1962. Concrete was poured day and night but was halted during the winter because of freezing temperatures. To organize the pouring, the dam was split into 45 ft (14 m) plots and each was raised about 5 or 6 ft. at a time. Workers had about 150 days - before seasonal flooding - to construct the dam to a height of at least 250 ft (76 m). Before flooding began, engineers planned to seal the diversion tunnels and begin filling the reservoir. To speed the pouring process, concrete was poured in casts by buckets that moved along three cableways suspended above the construction site. The deadline was met within the 150 days and a total of 1,000,000 cu yd (760,000 m3) of concrete was poured. The dam was eventually completed in 1968.

On September 25, 1968, the government-owned utility organized a ceremony to mark the completion of the Manicouagan-5 dam. Hundreds of dignitaries, politicians, utility executives, financiers, engineers and journalists were ferried by plane from Montreal, Quebec City and New York to the work-site to attend a banquet and a plaque unveiling ceremony.

Among the guests were Quebec Union Nationale Premier, Daniel Johnson Sr., his predecessor, Jean Lesage, and René Lévesque, the former Hydraulic Resources minister responsible for the consolidation of all investor-owned utilities into Hydro-Québec. Photographs taken at the banquet show the three men were in excellent spirits, holding hands and smiling, although relations between the Liberal leader and his former cabinet minister were strained by Lévesque's recent defection to the Mouvement Souveraineté-Association, a precursor of the Parti Québécois.

In his authorized biography, Hydro-Québec executive, Robert A. Boyd, recalls being woken up at 6 a.m. the next morning by his boss, Roland Giroux. "I've got bad news, Robert...", said Giroux, adding that he just found the Premier lying dead in his bed. Johnson's sudden demise sent shock waves at the work-site and across the province and the dedication ceremony was quickly canceled.

On September 26, 1969, a year to the day after Johnson's death, the new Premier Jean-Jacques Bertrand accompanied by Johnson's widow and children, unveiled two plaques and officially dedicated the dam after his predecessor. Both plaques now sit side by side at the top of the complex.

Facilities

Dam

The design was chosen for strength and economical reasons as it used less concrete or material than a gravity or embankment dam. The Daniel-Johnson Dam is a 214 m (702 ft) tall and 1,312 m (4,304 ft) long multiple-arch buttress dam. Of the dam's 14 total buttresses, the two that form the center arch are 530 ft (160 m) apart at their base while the others are about 250 ft (76 m) apart. At its thickest point, the center, the dam is 22.5 m (74 ft) wide while the crest can reach about 3 m (10 ft) wide. Pressure from water behind the dam is transferred from the dam's arches to its buttresses and lastly into the ground or its foundation.

The dam was constructed with a high-quality concrete designed to withstand constant thawing and freezing associated with its environment. To further help the structure cope with the climate, engineers placed one inch steel reinforcing bars within the upstream and downstream faces of the dam. The concrete's strength in compression was initially 4,500 lbs. per square inch to meet an estimated 1,500 PSI within the structure. The upstream face of the dam was also coated in asphalt for protection against water. Despite the strength of the dam's concrete, two parallel inclined cracks were discovered on one of the arches shortly after construction.

Reservoir

The dam impounded the Manicougan River, which filled the Earth's fifth largest confirmed impact crater, the Manicouagan crater, creating Manicouagan Reservoir. Coincidentally, the reservoir itself is the fifth largest in the world as well. The reservoir has a maximum depth of 350 m (1,150 ft) feet, mean depth of 85 m (279 ft) and contains 142 km3 (34 cu mi) of water. While draining an area of 29,241 km (18,170 mi), it has a surface area of 1,950 km2 (750 sq mi) and shoreline of 1,322 km (821 mi).

The reservoir is a well known area for Atlantic salmon, lake trout and northern pike fishing, although tall trees flooded during the impoundment have not decomposed due to a lack of oxygen, which can sometimes interfere with the sport.

The priming of the reservoir also created a large artificial island in the center of the Manicouagan reservoir by merging two crescent-shaped lakes: Mouchalagane Lake on the western side and Manicouagan Lake on the eastern side. Covering an area of 2,020 km2 (780 sq mi), René-Levasseur Island is considered to be the second largest island in the world located in a lake, in terms of area (the largest is Manitoulin Island in Lake Huron).

The island was named after René Levasseur, the chief engineer responsible for the construction of the Daniel-Johnson dam. Levasseur died at the age of 35, only days before the dam's inauguration.

Powerhouses

The dam fuels two powerhouses, the Manic-5 and Manic-5-PA. The first powerhouse consists of eight Francis turbines, capable of producing up to 1,528 MW of power, which went online in 1970. The second powerhouse, the Manic-5-PA (PA stands for puissance additionnelle or additional power), was commissioned in 1989, and consists of four Francis turbines of 1,064 MW in total installed capacity.

The designers of the Manic-5 decided on an above ground power house that was downstream of the dam for safety and cost. The intake was built on the east side of the dam and supplies two 3,400 ft (1,000 m), long 36 ft (11 m) diameter concrete-lined penstocks (tunnels). Just before reaching the power house and its eight turbines, each penstock splits into four branches.

The power house is about 2,500 ft (760 m) downstream of the dam and utilizes two surge tanks for sudden rises in water pressure from the two penstocks. Each surge tank has a 80 ft (24 m) diameter expansion chamber and is about 40 ft (12 m) higher than the actual dam structure. The surge tanks protect the penstocks and turbines from water hammer which would occur if the turbine gates were quickly closed and water pressure suddenly increases.

CULTURAL AND POLITICAL SIGNIFICANCE

The construction of the Daniel-Johnson Dam and the Manic-Outardes complex happened in a larger social and political context of the Quiet Revolution in Quebec, a time when recently nationalized "Hydro-Québec rapidly becomes a symbol of the new Quebec nationalism and of the new economic strategy of the State", explains historian Paul-André Linteau.

This newfound euphoric attitude is found in newspapers of the period. In a series of papers published in Montreal's La Presse, Renaude Lapointe calls Hydro-Québec a "colossus on the march". In this context, the construction of the complex is closely followed by the public and becomes part of the popular culture of that era.

For instance, chansonnier Georges Dor penned his huge 1966 hit, La Manic. The song tells the story of a construction worker at the remote jobsite who describes his loneliness to his wife in words that captured the collective imagination of the Quebec public.

Belgian-born novelist Henri Vernes was also inspired by the gigantic project and made the Manic-5 project the setting of one of his Bob Morane adventure novels. Terreur à la Manicouagan is about an attempt by arch-villain Roman Orgonetz to breach the dam by destroying the surge chamber by remote-controlled detonation, a plan foiled by Morane and his Scottish sidekick, Bill Ballantine. Prior to writing his novel, Vernes spent some time at the worksite, at the invitation of the Quebec government and Hydro-Québec. The book was launched in 1965 at the Hydro-Québec Building in Montreal.

Other notable visitors to the construction site included famous cartoonist Hergé, who left an original drawing of his characters Tintin the reporter and his dog Snowy posing in front of a ligne claire depiction of the dam. A colorized version of the autographed drawing is featured at the Jean-Lesage generating station visitor's center, 23 kilometres (14 mi) north of Baie-Comeau.

In the late 60s, a Montreal-based public relations officer with Renault Canada, Jacques About, was asked by the company to study the feasibility of introducing the Renault Alpine sports car into Canada. About's survey showed some potential but the manufacturer ultimately decided against introducing the model in its North American offerings.

The Montrealer decided to leave the company and start his own manufacturing business, Automobiles Manic Inc., in 1968. After gaining financial support from public and private backers, including Bombardier and Steinberg's supermarkets, his company started building a 2-seat sport coupe based on the Alpine A110, the Manic GT. Due to disappointing sales, the company stopped making the car in 1971.

The hydroelectric project also gave its name to Montreal's first foray into professional soccer, the short-lived Montreal Manic. The franchise competed for three seasons in the North American Soccer League in the early 80s. In March 2000, the Canada Post Corporation issued a 46 cent stamp featuring the Dam as part of a 4-stamp Millenium Collection sheet depicting "engineering and technological marvels".

Guided tours

The dam is also a tourist attraction. Since the 1960s, Hydro-Québec organizes four daily tours of the facility between June 24, Quebec's National Day, and August 31. Off-season tours can also be arranged by appointment. The 2-hour tour includes a briefing on the construction and operation of the facility, a visit of the Manic-5 powerhouse and a bus ride to the base and the crest of the Daniel-Johnson dam, which offers a spectacular panorama of the Manicouagan River valley. A belvedere, accessible by car 5 minutes away, is a nice picnic spot and offers the visitor a majestic view of the dam, which is lit at night .

Although picture-taking is allowed in the visitor's center and outside, cameras are forbidden inside the powerhouses, as part of tightened security measures implemented after a Radio-Canada television crew entered the unguarded Manic-5-PA and Robert-Bourassa powerhouses in February 2005. The incident led to the resignations of Hydro-Québec's CEO André Caillé, and chairman of the Board André Bourbeau, less than two months later. To access the dam from Baie-Comeau, visitors must take the narrow, winding Route 389. The dam is about 3 hours away. In 2009, 8,217 people visited this remote facility.

BLIGH'S CREEP THEORY

Design of impervious floor for sub surface flow: Iit is directly depended on the possibilities of percolation in porous soil on which the floor (apron) is built. Water from upstream percolates and creeps (or travel) slowly through weir base and the subsoil below it. The head lostby the creeping water is proportional to the distance it travels (creep length)along the base of the weir profile. The creep length must be made as big as possible so as to prevent the piping action. This can be achieved by providing deep vertical cut-offs or sheet piles.

According to Bligh's theory, the total creep length for first drawing: L = B and for second drawing: L = B + 2(d1 + d2 + d3)

If H is the total loss of head, then the loss of head per unit length of the creep shall be

$$C' = \frac{H}{L} = \frac{H}{[B + 2(d_1 + d_2 + d_3)]}$$

Bligh called the loss of head per unit length of creep as Percolation coefficient. The reciprocal, (L/H) of the percolation coefficient is known as the coefficient of creep C.

Assumptions

- Hydraulic slope or gradient is constant throughout the impervious length of the apron.

- The percolating water creep along the contact of the base profile of the apron with the sub soil losing head enroute, proportional to length of its travel. The length is called creep length. It is the sum of horizontal and vertical creep.
- Stoppage of percolation by cut off (pile) possible only if it extends up to impermeable soil strata.

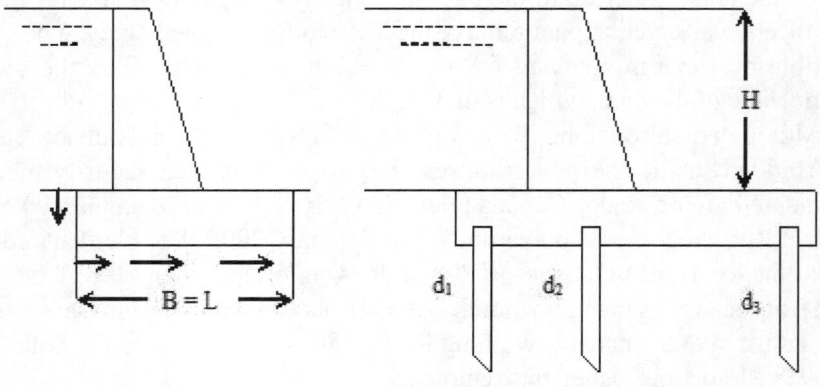

Method of increasing creep length

Design cr iteria

Safety against piping:

The length of creep should be sufficient to provide safe hydraulic gradient according to the type of soil.

Safe creep length = L = CH, C = 1/c

Safety against uplift pressure:

Let h' = uplift pressure head at any point of apron (Hydraulic gradient line above the bottom of floor)

The uplift pressure = wh' where w = density of water. If t = thickness of floor at the point, l = specific gravity for floor material. Then, downward force per area (resisting force) = t.w.e or wh' = t.w.e

For portion of floor upstream of barrier only nominal thickness need to be provided since the weight of water will counterbalance the uplift pressure.

A certain minimum length of impervious floor is always necessary to the downstream of the barrier (thickness of downstream floor for worst condition)

Limitations of Bligh's theory

1. This theory made no distinction between horizontal and vertical creep.
2. Did not explain the idea of exit gradient - safety against undermining cannot simply be obtained by considering a flat average gradient but by keeping this gradient will be low critical.
3. No distinction between outer and inner faces of sheet piles or the intermediate sheet piles, whereas from investigation it is clear, that the outer faces of the end

sheet piles are much more effective than inner ones.

4. Losses of head does not take place in the same proportions as the creep length. Also the uplift pressure distribution is not linear but follow a sine curve.
5. In case of two piles the width between should be greater than twice the head or the piles are not effective.

Khosla's theory

After studying a lot of dam failure constructed based on Bligh's theory, Khosla came out with the following;

Following are some of the main points from Khosla's Theory

- From observation of Siphons designed on Bligh's theory, by actual measurement of pressure, with the help of pipes inserted in the floor of two of the siphons?
- Does not show any relationship with pressure calculated on Bligh's theory. This led to the following provisional conclusions:
- Outer faces of end sheet piles were much more effective than the inner ones and the horizontal length of the floor.
- Intermediated piles of smaller length were ineffective except for local redistribution of pressure.
- Undermining of floor started from tail end.

For case of two dimensional flows under a straight floor where:

Thus for the first flow line AB which touches the outline of the floor, the pressure can be obtained by putting different values of x in equation. Fig shows the pressure distribution diameter both by equation 4 as well as Bligh's Theory.

From the fig the following conclusions can be drawn:

Slope of Pressure diagram: At A and B in infinite, hence the floor at A will be theoretically infinite acting downward and that at B will also be infinite acting upward. This will cause sand boiling and hence the floor should be depressed or cut off should be provided at the downstream end.

Composite profile:

The following specific causes of general form were considered.

- Straight horizontal flow of negligible thickness with pile at either end, upstream or at downstream end.
- Straight horizontal floor of negligible thickness with pile at some intermediate point.
- Straight horizontal floor, depressed below the bed, but with no cut off.

3Method of independent variable:

- Most designs do not confirm to elementary profiles (specific cases). In actual cases we may have a number of piles at upstream level, downstream level and intermediate points and the floor also has some thickness.
- Khosla solved the actual problem by an empirical method known as method of

independent variables.

- This method consists of breaking up a complex profile into a number of simple profiles, each of which is independently amiable to mathematical treatment. Then apply corrections due to thickness of slope of floor.
- As an example the complex profile shown in fig is broken up to the following simple profile and the pressure at Key Points obtained.
- Straight floor of negligible thickness with pile at upstream ends.
- Straight floor of negligible thickness with pile at downstream end.
- Straight floor of negligible thickness with pile at intermediate points.
- The pressure is obtained at the key points by considering the simple profile.

For the determination of seepage below the foundation of hydraulic structure developed the method of independent variable.

In this method, the actual profile of a weir which is complex, is divided into a number simple profiles, each of which cab be solved mathematically without much difficulty. The most useful profile considered are:

A straight horizontal floor of negligible thickness provided with a sheet pile at the upstream end or a sheet pile at the downstream end.

ii) A straight horizontal floor depressed below the bed, but without any vertical cut-off.

1. A straight horizontal floor of negligible thickness with a sheet pile at some Intermediate point

The mathematical solution of the flow-nets of the above profiles have been given in the form of curves. From the curves, percentage pressures at various key points E, C, E1, C1 etc) be determined. The important points to note are:

1. Junctions of pile with the floor on either side{E, C (bottom), E1, C1 (top) }
2. Bottom point of the pile (D), and
3. Junction of the bottom corners (D, D') in case of depressed floor

The percentage pressures at the key points of a simple forms will become valid for any complex profile, provided the following corrections are effected:

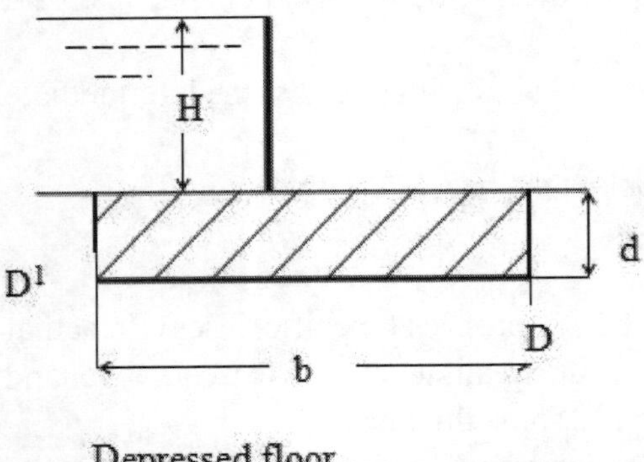

Depressed floor

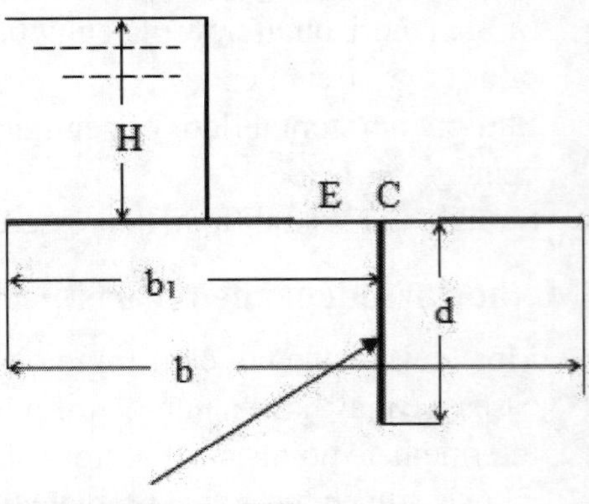

Sheet pile at intermediate position

- correction for mutual interference of piles
- correction for the thickness of floor
- correction for slope of the floor.

Correction for Mutual Interference of Piles

Let b1 = distance between the two piles 1 and 2, and

D = the depth of the pile line (2), the influence of which on the neighbouring pile (1) of depth d must be determined

b = total length of the impervious floor

c = correction due to interference.

The correction is applied as a percentage of the head

$$C = 19\sqrt{\frac{D}{b_1}}\left(\frac{d+D}{b}\right)$$

This correction is positive when the point is considered to be at the rear of the interfering pile and negative for points considered in the forward or flow direction with the interfering pile.

Correction for Floor Thickness

Standard profiles assuming the floors as having negligible thickness. Hence the values of the percentage pressures computed from the curves corresponds to the top levels (E1*, C1*) of the floor. However, the junction points of the floor and pile are at the bottom of the floor (E1, C1)

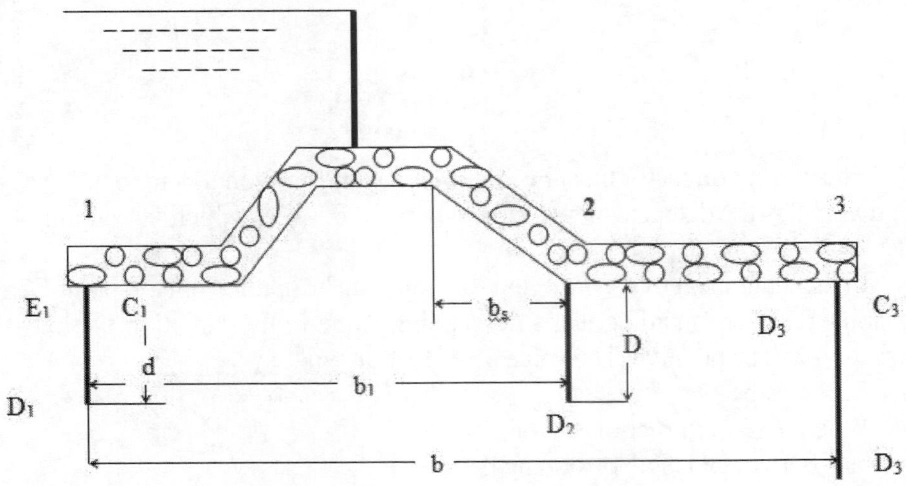

The pressures at the actual points E1 and C1 are interpolated by assuming a straight line variation in pressures from the points E1* to D1 and from D1 to C1

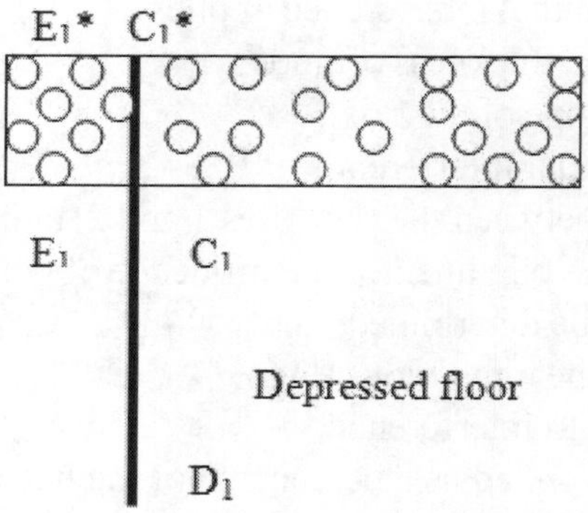

The corrected pressures at E1 should be less than the computed pressure t E1*. Therefore the correction for the pressure at E1 will be negative. And so also is for pressure at C1.

Correction for Slope of Floor

A correction for a sloping impervious floor is positive for the down slope in the flow direction and negative for the up slope in the direction of flow.

No.	Slope = Ver:Horiz	Correction as % of pressure
1	1:1	11.2
2	1:2	6.5
3	1:3	4.5
4	1:4	3.3
5	1:5	2.8
6	1:6	2.5
7	1:7	2.3
8	1:8	2.0

The correction factor must be multiplied by the horizontal length of the slope and divided by the distance between the two poles between which the sloping floor exists.

In the diagram above, correction for slope can be applied only to point E2. As the point E2 is terminating at the descending slope in the direction of flow, the correction will be positive. The value of correction will be:

C.F. x bs/b1

Where C.F. =correction factor

bs = horizontal length of sloping floor

b1 = horizontal distance between the pile lines

7

Watershed Management

Watershed management is the study of the relevant characteristics of a watershed aimed at the sustainable distribution of its resources and the process of creating and implementing plans, programs, and projects to sustain and enhance watershed functions that affect the plant, animal, and human communities within the watershed boundary. Features of a watershed that agencies seek to manage include water supply, water quality, drainage, stormwater runoff, water rights, and the overall planning and utilization of watersheds. Landowners, land use agencies, stormwater management experts, environmental specialists, water use surveyors and communities all play an integral part in watershed management.

CONCEPTS OF WATERSHED MANAGEMENT

Watershed management uses a holistic management approach of soil, water, and vegetation in the watershed area in order to increase agricultural productivity and conserve natural resources. The goal is to improve peoples' standard of living without damaging the environment. Improving management of watersheds requires investments in technology, infrastructure, social institutions and markets. Typical watershed management technologies include soil conservation, water harvesting, storage and distribution, increasing vegetative cover, and safe disposal of excess water. International experience shows that successful watershed management requires that these technologies be adapted to local climatic and topographical conditions; population densities; crop types and agricultural practices; institutional capacities; and it must ensure that poor people benefit from it. Consequently, investments are also necessary to build the capacity of local institutions and communities. Farming systems improvements and value addition/marketing are key ingredients for deepening the impact of watershed investments.

Many countries are now trying to place watershed management within the context of natural and human systems (Bunch et al. 2014; Sanchez et al. 2014). The World Bank and the United Nations Food and Agriculture Organization (FAO), amongst others, use watershed management approaches to assess the environmental benefits of development projects and aim to improve the provision of goods and services from watersheds including agriculture, forestry, and fisheries in a sustainable manner (Kumar et al. 2014; Sha et al. 2014). Components of watersheds such as low-lying lakes can serve as indicators for overall ecosystem health and change as they aggregate materials from the water and air (Hampton 2013) and thus are an appropriate means to assess the larger system. Watershed management recognizes that people are affected by the interaction of water with other resources and that people can influence the nature and magnitude of those interactions (Manuelli et al. 2014).

In South Asia, the World Bank has provided assistance for watershed projects in rainfed areas for more than 20 years with demonstrated successful impacts. In India, for example, the Integrated Water Development Hills II project has increased agricultural productivity and incomes of beneficiaries while sustaining the natural resource base in fragile and degraded agro-ecosystems. The Karnataka Watershed Development project has developed new approaches for monitoring and evaluation linked to remote sensing and Geographic Information Systems (GIS). The recently launched Uttaranchal Decentralized Watershed Development project is developing the capacity of local government institutions to play a more meaningful role in planning, implementing, and managing watershed and other economic activities in rural areas.

Watershed Management Important

Runoff from rainwater or snowmelt can contribute significant amounts of pollution into the lake or river. Watershed management helps to control pollution of the water and other natural resources in the watershed by identifying the different kinds of pollution present in the watershed and how those pollutants are transported, and recommending ways to reduce or eliminate those pollution sources.

All activities that occur within a watershed will somehow affect that watershed's natural resources and water quality. New land development, runoff from already-developed areas, agricultural activities, and household activities such as gardening/lawn care, septic system use/maintenance, water diversion and car maintenance all can affect the quality of the resources within a watershed. Watershed management planning comprehensively identifies those activities that affect the health of the watershed and makes recommendations to properly address them so that adverse impacts from pollution are reduced. Watershed management is also important because the planning process results in a partnership among all affected parties in the watershed. That partnership is essential to the successful management of the land and water resources in the watershed since all partners have a stake in the health of the watershed. It is also an efficient way to prioritize the implementation of watershed management plans in times when resources may be limited.

Because watershed boundaries do not coincide with political boundaries, the actions of adjacent municipalities upstream can have as much of an impact on the downstream municipality's land and water resources as those actions carried out locally. Impacts from upstream sources can sometimes undermine the efforts of downstream municipalities to control pollution. Comprehensive planning for the resources within the entire watershed, with participation and commitment from all municipalities in the watershed, is critical to protecting the health of the watershed's resources.

Types of Watershed Management

Watershed is classified depending upon the size, drainage, shape and land use pattern.

a. Macro watershed: 1000 -10,000 ha
b. Micro watershed: 100 -1000 ha
c. Mini watershed: 10 -100 ha
d. Mille watershed: 1 -10 ha

Watersheds for Sustainable Resource Management

Watershed management is an adaptive, comprehensive, integrated multi-resource management planning process that seeks to balance healthy ecological, economic, and cultural/social conditions within a watershed. Watershed management serves to integrate planning for land and water; it takes into account both ground and surface water flow, recognizing and planning for the interaction of water, plants, animals and human land use found within the physical boundaries of a watershed.

Watershed management provides a framework for integrated decision-making to help: assess the nature and status of the watershed; identify watershed issues; define and re-evaluate short and long-term objectives, actions and goals; assess benefits and costs; and implement and evaluate actions.

Adopting a watershed approach is founded on the basis that Alberta's water resources must be managed within the capacity of individual watersheds and that all Albertans recognize there are limits to the available water supply. What happens on the land and water in a watershed can affect the water supply that rivers provide. While land and water are closely linked, these resources have not historically been managed in a fully integrated manner. Focusing efforts at the watershed level provides a comprehensive understanding of local management needs, and encourages locally led management decisions.

Watersheds are natural environmental and land management units that determine the health of a nation. Poor ecosystem management within watersheds has and will result in the impaired functioning of the watershed, which in fragile environments can lead to ecosystem collapse (Eswaran and Samra 1997).

Watershed management has moved from a focus on physical water and soil utilization and conservation to the integration of social, economic, and environmental development.

As part of these changes, traditional watershed management must shift its conventional focus away from wildlands to include the urban fringe and urbanized areas if it is to satisfy society's needs (Nearly 2000). Watershed management must also be capable of providing solutions to watershed problems such as plans for water augmentation or watershed restoration. That being said, watershed management can be difficult to regulate; while regulations can prevent people from undertaking harmful actions, the regulations often provide no incentive for people to take beneficial actions. As such, watershed management requires the application of the three "Ps": planning, partnership, and participation by stakeholders (Blomquist and Schlager 2005). Public-private partnerships (PPPs) are becoming a more prevalent way to improve management success; PPPs join these two sectors to work towards a common goal and enables each to benefit from the other's strengths and resources to more effectively meet management objectives (PPPIRC 2016). Additionally, watershed management relies heavily on the science of watershed (forest/range/wildland/land use) hydrology, a branch of hydrology that addresses the effects of vegetation and land management on water quality, erosion, and sedimentation. Embedded in both watershed hydrology and management is the acknowledgment of the linkages between upstream and downstream areas and interrelationships among land use, soil, and water. With the increasing awareness that land management decisions cannot be made in isolation, the principles of watershed management are being used as the basis for many environmental and natural resource management decisions (Hebin and Ueta 2012; Karcher et al. 2013).

MICROWATERSHED MANAGEMENT

The rampant growth of population and advancements in life style have tremendously increased the demands for food, fuel, fodder, fiber, shelter, communication, etc. These growing demands are putting the resilience of the natural resource base under threat. To ensure food and water security, the vertical and horizontal expansion of production, has to be effective without degrading productivity.

Development through watershed approach is one such developmental option. Watershed management is the study of the relevant characteristics of a watershed, aimed at the sustainable distribution of its resources and the process of creating and implementing plans, programmes and projects to sustain and enhance watershed functions that affect the plant, animal and human communities within a watershed boundary. Watershed is a hydrological unit bounded by natural ridges and allows the run off due to rainfall to drain in a well defined drainage pattern of streams flowing within the watersheds boundary. It aims at alleviating habitat and inhabitant impoverishment through a holistic approach of conservation and sustainable exploitation of natural responsibility and harmonious co-existence manner.

A watershed is an area of land and water bounded by a drainage divide within which the surface runoff collects and flows out of the watershed through a single outlet into a lager river (or) lake. Land and water are the two most valuable and essential resources which forms the basis of all the life and forms key resources in all economic activities ranging from agriculture to industry. Watershed management is no longer the exclusive domain of soil and water conservationists, it now needs equal contribution from social scientists, doctors and experts of other disciplines also.

As an integral part of area development programme, land resources development programmes are taken upon micro watershed basis as the shape of the watershed controls, the natural resources like water, soil and productivity of the land for successful implementation of agriculture, forest and other micro level development in each hectare of a watershed the micro watershed and the village becomes the most adaptable unit. For regional planning and management the micro watersheds and villages are the most workable units. Using satellite data and GIS technology, maps of landuse/landcover, drainage, soil, etc., will became indispensable to manage natural and human resources.

The concept of development of land and water resources on watersheds basis gained importance in India since 1974. The watershed approach is increasingly being deployed in various development programmes to manage the water and land resources like soil and water conservation.

Integrated watershed management grapples with the complexity of interactions between ecosystems and socio-economic systems, and aims to restore and sustain the health, productivity, and biodiversity of ecosystems through strategies that integrate the needs of society and the economy (Szaro et al. 1998; Einar 2010; Qi and Altinakar 2013). The concept of integrated watershed management is fairly easy to envisage, but the practice of integration is complicated. It is a dynamic process that crosses temporal-spatial spectra, jurisdictional boundaries, and social, cultural, economic, and environmental systems. In order to succeed, it must be participatory, adaptive, and experimental, involve all pertinent stakeholders, identify an appropriate balance between development and protection, and integrate all relevant scientific knowledge and user-supplied information about the social, economic, and environmental processes affecting natural resources within the watershed (Calder 1999; Yang et al. 2006; VanHouten 2014). Additionally, effective integration is largely dependent on current regional development levels and the aims of future development (Thorburn 2012; Mutekanga et al. 2013; Ozturk et al. 2013). Through monitoring of and research on ecological and socio-economic interactions and processes, integrated watershed management can remain adaptable and help to develop strategies to sustain ecosystem composition, structure, and function through policies, protocols, and practices that are based on sound research.

Adaptive management (AM) can be coupled with integrated management and can improve the ability to cope with the inherent uncertainties of managing complex, dynamic systems such as watersheds by learning from the outcomes of management implementation and adjusting future approaches accordingly (Allan et al. 2008; Porzecanski et al. 2012). For AM to be successful, it requires set timeframes for management evaluation where failures and inadequacies can be openly discussed and dealt with. Management strategies need to be continuously improved by learning from the implemented policies and remaining flexible (Raadgever et al. 2008). This requires financial support as well as the dedication of time and resolution of those involved to critically and objectively evaluate their own work (Allan et al. 2008). Without the dedication and means to identify mistakes and rectify them, AM will not be successful.

The use of Traditional Ecological Knowledge (TEK) is also a key component in successful watershed management and can be incorporated into adaptive and integrated management techniques. It is the accumulation of knowledge, experience, and values regarding the local ecosystem held by communities with a history of subsistence living that addresses the interactions among and between organisms (including humans) and their environment (Berkes et al. 2000; Olsson and Folke 2001; Ellis 2005). Traditional knowledge approaches are holistic and adaptive by nature, acknowledge ever changing environmental conditions, and are often derived with the purpose of maintaining ecosystem structure and function while providing resources to the local community-a goal similar to those of today's watershed management strategies. There are several components of TEK that may be complementary to adaptive management techniques, such as management regulations that are locally developed and enforced by the users; the flexible use of resources through rotations; innate flexibility within the approach of decision-makers to respond to environmental feedbacks; the use of an array of resources to minimize environmental risk; and the use of qualitative measurements to evaluate the direction of future management strategies (Berkes et al. 2000). However, there are several barriers to the successful incorporation of TEK into modern management strategies. These include conceptual barriers due to differing values, communication barriers arising from different languages, and political barriers resulting from the unwillingness to acknowledge traditional knowledge that opposes political or industry agendas (Ellis 2005). For instance, certain traditional practices may be preferentially selected because they conform more to Western-style governance and values. To integrate TEK in watershed management more successfully, there needs to be a shift from forcing selected elements of traditional practices into conventional strategies to making current methods more malleable so that they may be integrated with TEK.

In general, the watershed management process has 6 steps (Fig. 1): (1) Survey the status of the watershed and identify its situation; (2) identify stakeholders; (3) identify interests and objectives; (4) determine the target and plan; (5) implement the plan; and (6) evaluate management success and failures, reassess objectives, and adjust the plan to improve management success. In the first step, the implementing agency, whether it is private, governmental, academic, etc., needs to survey indicators of ecosystem health to evaluate ecosystem function, such as physico-chemical (e.g., pH, water temperature, concentration of nutrients), biological (e.g., flora and fauna biodiversity, algal growth), habitat (e.g., riparian habitat species composition, degree of bank erosion), and water flow indicators (e.g., peak flow, base flow) (Queensland Government 2016). This step will help identify any ecological issues and trends, which will help inform the decision-making and planning process. Next, stakeholders need to be identified; this includes communities, people, and organizations that have an interest in the watershed and who will be influenced by management decisions and outcomes.

The third step is to identify stakeholders' interests and management objectives. This information can be obtained through surveying and engagement with stakeholders and from the initial ecosystem surveys. The next step is to determine the management target and plan based on the information gathered during the ecological survey and the concerns and needs of stakeholders. For the fifth step, the management plan needs to be implemented by the responsible agency; they should oversee implementation, monitoring, and enforcement of management strategies. Lastly, throughout the life of the management plan, the implementing agency needs to assess and improve their management plan to account for changes in the environment and stakeholder needs and address management activities that are insufficiently meeting management objectives.

This self-assessment and evaluation involves reflection and evaluation of the effectiveness of the management plan in meeting its objectives; development of solutions to improve management in areas where objectives are not being met; identification of new social or ecological issues related to the watershed; adjustment of management objectives where needed to more effectively meet existing and new management issues; and redevelopment of the management action plan to reflect the new and improved objectives that account for lessons learned in the self-assessment of management success. Developing site-specific management plans based on these generalized steps can lead to watershed management strategies that achieve the goals of sustainability in relation to land resource use, the ecosystem, the ecological economy, and human health and well-being.

Open image in new window

In summary, watershed management provides a framework to integrate decision-making to help assess the nature and status of the watershed; identify watershed issues; define and re-evaluate short- and long-term objectives, actions and goals; assess benefits and costs; and implement and evaluate actions. Sustainable utilization and management of resources and the environment is the key to watershed management (Wagner et al. 2002; Sungjun 2007; Davenport and Seekamp 2013). Additionally, an equitable partnership between stakeholders is a crucial component to watershed management (Davenport and Seekamp 2013; Mutekanga et al. 2013; Manuelli et al. 2014). Now, sustainable watershed management has become a common process, and sustainability guidelines on watershed management are available for industrialized, newly industrialized, and developing countries to generate better, more locally specific, watershed management strategies.

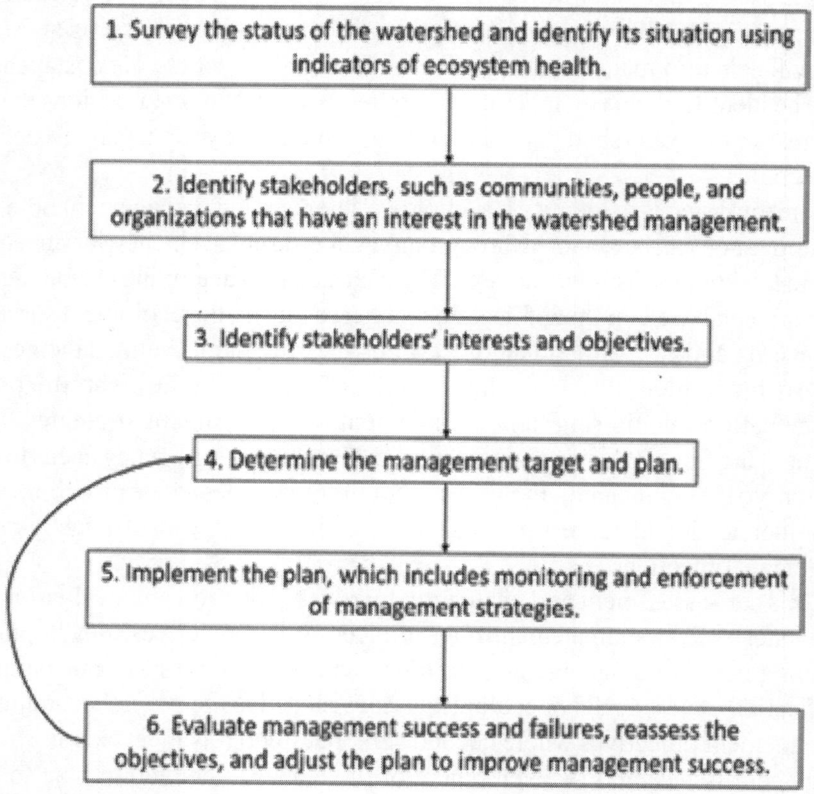

Fig.: Conceptual model for developing an integrated watershed management plan

Steps in watershed management

Watershed management involves determination of alternative land treatment measures for, which information about problems of land, soil, water and vegetation in the watershed is essential. In order to have a practical solution to above problem it is necessary to go through four phases for a full scale watershed management.

Programme

1. Recognition phase.
2. Restoration phase.
3. Protection phase.
4. Improvement phase.

1. Recognition Phase

It involves following steps

1. Recognition of the problem
2. Analysis of the cause of the problem and its effect.
3. Development of alternative solutions of problem.

Necessary information is obtained from different surveys like soil survey, land capability survey, agronomic survey, forest, engineering and socio economic survey, etc. This information serves as a basis for fixing and determining the watershed problems, priorities in land treatment measures, and causes and effects of problems on land and people.

2. Restoration Phase

It includes two main steps.

a. Selection of best solution to problems identified

b. Application of the solution to the problems of the land

As per the priorities, treatment applied initially to critical areas. After this proper measures like biological and engineering measures are applied to all types of lands.

3. Protection Phase

This phase takes care of the general health of the watershed and ensures normal functioning. The protection is against all factors which may cause determined in watershed condition.

4. Improvement phase

This phase deals with overall improvement in the watershed and all land is covered. Attention is paid to agriculture and forest management and production, forage production and pasture management, socio economic conditions to achieve the objectives of watershed management. Health, family planning, improving cattle, poultry, etc. are taken depending upon intensity.

Familiarize Yourself with Your Watershed

Comprehensive watershed plans should first identify the characteristics of the watershed and inventory the watershed's natural resources. It is important to establish a baseline of the overall nature and quality of the watershed in order to plan properly for the improvement of the resources in the watershed and to actually measure those improvements.

The first steps in watershed management planning are to:

- Delineate and map the watershed's boundaries and the smaller drainage basins within the watershed;
- Inventory and map the resources in the watershed;
- Inventory and map the natural and manmade drainage systems in the watershed;
- Inventory and map land use and land cover;
- Inventory and map soils;
- Identify areas of erosion, including stream banks and construction sites;

- Identify the quality of water resources in the watershed as a baseline; and
- Inventory and map pollution sources, both point sources (such as industrial discharge pipes) and nonpoint sources (such as municipal stormwater systems, failing septic systems, illicit discharges).

Much of this information may already be compiled and available through the DEP, the Natural Resources Conservation Service of the U.S. Department of Agriculture, and municipal offices such as planning and zoning, inland wetlands, and public works. Additional information specific to the watershed can be gathered during volunteer stream walks which allow for on the ground study of the general conditions of the receiving waters and the adjacent watershed areas.

BUILD LOCAL PARTNERSHIPS

Watershed planning should also identify and include the partners, or "stakeholders," in the watershed. Development of local partnerships can also lead to greater awareness and support from the general public. Once individuals become aware of and interested in their watershed, they often become more involved in decision-making as well as hands-on protection and restoration efforts. Through such involvement, watershed management builds a sense of community, helps reduce conflicts, increases commitment to the actions necessary to meet environmental goals, and ultimately, improves the likelihood of success for the watershed management plan. Local partnerships can include:

- Residents;
- Landowners;
- Federal, state, and municipal government officials;
- Watershed associations and other environmental and civic groups;
- Local business and industry leaders;
- Agricultural users;
- Developers;
- Teachers; and
- Recreational users.

DETERMINE PRIORITIES FOR ACTION

Watershed management planning should also determine what the opportunities are to reduce pollution or address other pressing environmental issues, prioritize those opportunities, and identify a time frame for accomplishing pollution reduction and resource and habitat improvements. Those issues that pose the greatest risk to human health or particular resources, or to desired uses of resources (i.e., swimming beaches), might be given highest priority for control and reduction. Watershed plans should establish clear goals, visions, and actions to be taken.

Examples of opportunities to reduce pollution and address other wide-ranging environmental issues include:

- Infrastructure improvements. More frequent maintenance of municipal stormwater systems or improving or replacing inadequate stormwater treatment systems, identifying and eliminating illicit (i.e., non-stormwater) connections to municipal stormwater systems;
- Reducing paved areas and other impervious cover, especially adjacent to waterbodies and wetlands. Zoning and subdivision regulations can be revised to address issues such as reducing lot coverage/impervious cover, reducing roadway widths, encouraging cluster and low impact development, limiting land disturbance such as grading and clearing, and increasing development setbacks from resources;
- Identifying appropriate areas for open space acquisition, greenways planning, and the establishment of vegetated buffers along waterbodies and wetland areas;
- Establishing sewer avoidance areas to limit development;
- Increasing inspections and maintenance of existing septic system and encouraging repairs to failing systems;
- Identifying other appropriate housekeeping practices for homeowners and landowners (encouraging the use of vegetated buffers adjacent to waterbodies and wetlands, reducing lawn areas and the amount of fertilizers and chemicals applied to them, recommending washing cars over lawns instead of driveways so rinse water can drain into the lawn and not run-off into storm drains, etc.);
- Identifying resource and wildlife habitat restoration priorities;
- Increasing and promoting public access and greenways and identifying areas where it is appropriate to do so; and
- Identifying and evaluating opportunities for nonstructural flood protection efforts;
- Improving waste management, pollution prevention, and recycling efforts at municipal facilities and businesses within the watershed.

CONDUCT EDUCATIONAL PROGRAMS

The degree of public education and participation in the planning process can greatly influence the success of watershed management. There are many ways to involve and educate the public in watershed management. The formation of citizen review groups and advisory committees can gain public support from the watershed and are an essential component to a successful, community-based, and locally led effort. These community-based groups and committees can also provide the means to keep the project going once the plan has been finalized to make sure that recommended actions are taken. It might also be helpful to identify a watershed coordinator to help in this effort.

- Outreach and education efforts can include:
- Periodic informational meetings;
- Stream walk assessments;
- Organized storm drain stenciling projects;
- Watershed clean-up days and riparian planting/habitat restoration days;

- Coordination with school systems within the watershed;
- Information kiosks and websites;
- Videos; and
- Newsletters and other printed materials to provide status and progress reports.

Ensure Implementation and Follow-up

It is important to establish a schedule with milestones and some sort of committee to ensure that projects proceed in a timely manner. A monitoring program should also be established to measure success through data gathering. It is also important to identify ways in which landowners can be assisted with undertaking necessary improvements, such as low interest loans or technical outreach information. Finally, it is important to ensure that the recommendations contained in the watershed plan, especially design standards, are integrated into municipal land use regulations (zoning, subdivision, inland wetlands).

Steps in Watershed planning

Watershed management planning is a process that results in a plan or a blueprint of how to best protect and improve the water quality and other natural resources in a watershed. Very often, watershed boundaries extend over political boundaries into adjacent municipalities and/or states. That is why a comprehensive planning process that involves all affected municipalities located in the watershed is essential to successful watershed management.

- Preparation of base maps for carrying out surveys.
- Reconnaissance survey of the watershed for overall development.
- Assessing rainfall characteristics.
- Preparation of soil maps and classification of lands for different uses according to capability classification for agriculture, forestry, pasture, horticulture, etc.
- Preparation of inventory of existing land uses and farm sizes.
- Appraisal of agricultural production patterns and potentials, present and potential markets and possible group action arrangements.
- Carrying out topographic and hydrologic surveys for engineering works.
- Geo-hydrological survey to delineate areas suitable for groundwater development.
- Formulation of an integrated time-bound plan for land and moisture conservation, ground water. recharge, development of productive afforestation, agriculture production, grasslands and horticulture.
- Assigning of priorities for implementation of the project.
- Assessing social costs and benefits.

LAND CAPABILITY CLASSIFICATION

Watershed management must take into account land capability classification, since a typical watershed may include lands suitable for cropping, trees, pastures, etc. The land classification is as follows:

Class I

This type of land is nearly level, with minimum erosion, the soils are deep, well drained, with adequate water holding capacity. These soils need ordinary management practices to maintain productivity. Such practices may include the use of one or more of the following fertilisers, lime cover and green manure crops, conservation of crop residues aAd crop rotation. They are fit for agriculture, pasture, forests, etc.

Class II

These soils have some limitations. They are subject to slight erosion, as they are slightly sloping lands, with less soil depth, occasional drainage problems that can be easily corrected, and they are sometimes saline. They may need one or more of the following practices terracing, strip-cropping, contour cultivation, excess water drainage, crop rotation, use of green manure crops, stubble mulching, the use of fertilisers, manure and lime.

These soils may be used for growing cultivated crops, raising pastures, forests, etc.

Class III

The factors that limit plant growth are more severe than in class II. These factors could be moderately sloping land, land susceptible to water or wind erosion, land more susceptible to drainage problems, land with water-logging, shallow soil depth, low moisture holding capacity and moderate salinity.

Class IV

The limitations to plant growth is more severe here than in Class III soils. Very careful management is required and the conservation practices are more difficult to apply and maintain. Some of the factors affecting these lands are steep slopes, severe susceptibility to water and wind erosion, shallow soils, low moisture holding capacity, severe water-logging and severe salinity. These soils can be used for crops, pastures, forests and wildlife food, etc.

Class V

Cultivation is not feasible because of one or more limitations, such as over flow, stoniness, wetness or severe climatic conditions. Examples of this class are soils of low lands subject to frequent over flows which prevent the normal production of cultivated crops, or rocky or stony terrain. These soils are suitable for raising cultivated crops but are not suitable for perennial vegetation.

Class VI

These soils have very severe limitations that make them generally unsuitable for cultivation and limit their use largely to pastures or forests or wildlife food and cover. Some of the limitations which cannot be corrected are: steep slope, very severe erosion hazard, shallow rooting zone, excessive wetness, low moisture capacity and severe salinity.

Class VII

The physical condition of the soil is such that it is not practical to adopt pasture improvements and water control practices.

Class VIII

Bad lands, rocky out crops, sandy beaches, marshes, deserts, mine tailings and other nearly barren lands may be included in this class. It may be necessary to give protection and management for plant growth to soils and land forms in class VIII in order to protect other more valuable soils, to control water, or for wildlife or for aesthetic reasons.

8

Water management in High Water Table Areas

The water table is the upper surface of the zone of saturation. The zone of saturation is where the pores and fractures of the ground are saturated with water.

The water table is the surface where the water pressure head is equal to the atmospheric pressure (where gauge pressure = 0). It may be visualized as the "surface" of the subsurface materials that are saturated with groundwater in a given vicinity.

Groundwater flow

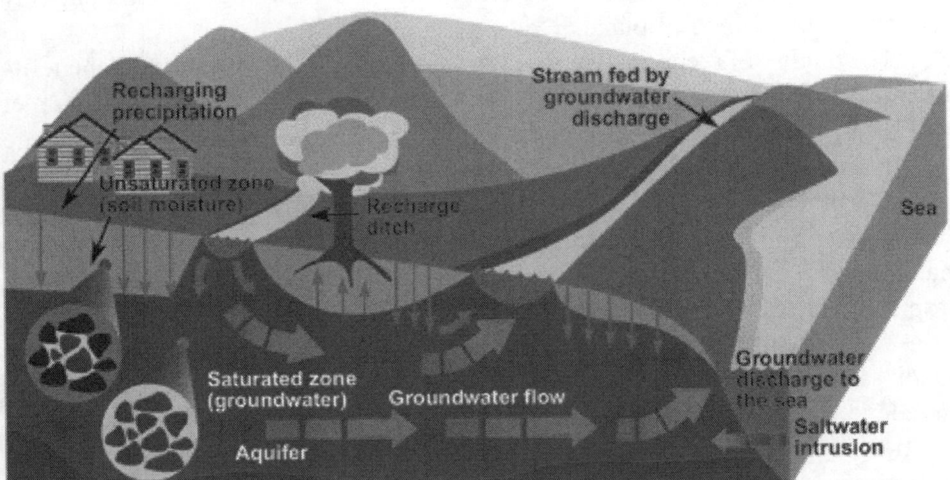

The groundwater may be from precipitation or from groundwater flowing into the aquifer. In areas with sufficient precipitation, water infiltrates through pore spaces in the soil, passing through the unsaturated zone. At increasing depths, water fills in more of the pore spaces in the soils, until a zone of saturation is reached. Below the water table, in the phreatic zone (zone of saturation), layers of permeable rock that yield groundwater are called aquifers. In less permeable soils, such as tight bedrock formations and historic lakebed deposits, the water table may be more difficult to define.

The water table should not be confused with the water level in a deeper well. If a deeper aquifer has a lower permeable unit that confines the upward flow, then the water level in this aquifer may rise to a level that is greater or less than the elevation of the actual water table. The elevation of the water in this deeper well is dependent upon the pressure in the deeper aquifer and is referred to as the potentiometric surface, not the water table.

A water table - sometimes also referred to as a groundwater table - is the upper level of accumulated water that has collected on the saturated or impermeable surface below it. Groundwater tends to level out just like the water in your bath or swimming pool. That's why it's called a table by the experts, because the tops of tables are usually smooth too.

There are some agronomic and water management practices which may be implemented at the field and watershed scales to reduce the concentration of chemicals (primarily nutrients and pesticides) in agricultural drainage water. Agronomic practices include tillage, cropping methods, and improved and timely fertilizer and pesticide applications.

One water management practice which has been shown to significantly reduce NO3-N in drainage water is water table control. The technique has been tested in the humid regions of eastern Canada, and the eastern and mid-western United States. It is currently promoted in these regions as a combined drainage/irrigation practice for increasing crop yields and reducing NO3-N pollution from relatively flat, artificially drained cropland.

The quality of receiving waters is thus improved. The water quality benefits of water table management are well described in Gilliam (1987) and Gilliam et al., (1979). Water table management systems are currently being tested in some parts of Europe and Asia.

The practice is yet to be evaluated in the arid and semi-arid regions, although there is potential for water table management in these regions. However, proper management of the water table is required to ensure that re-salinization of the crop root zone does not occur.

Early attempts at sub-irrigation in arid and semi-arid regions failed because of soil salinization. More information is required on the effects of different water table levels on crop yields and salt movement in the root zone in various climatic locations and soil types. For these reasons, the focus of this chapter is on water table management in humid regions.

In the humid regions of North America, water table management has evolved on lands which were first improved with surface and subsurface drainage. Drainage improvements were initially required on these poorly drained soils to increase crop productivity and provide better soil conditions for field machine trafficability. With concerns about the quality and impacts of agricultural drainage water on marine ecosystems, as well as nitrate contamination of drinking water, the drainage systems were modified to reduce the concentration of pollutants in the drainage water. Water table management has been practised for several years on organic and sandy soils in humid regions to reduce drought stress of high value crops.

Two Types of Water Tables

Geologists speak of two kinds of water tables, namely perched ones and true ones. True tables are those prevailing in an area. They represent the average water height, plus / minus the prevailing season in terms of wet or dry. However water doesn't always seep down consistently. Factors that affect water penetration include:

- The degree of existing saturation
- The permeability of the strata
- The presence of impermeable rock.

Groundwater may perch higher at some points than others as a result of this phenomenon. Think of this as a step-change, or as if there's a thick book laying on your desk or table.

Implications for Building Basements

Homeowners watching their new basement being excavated should not necessarily be content when the earth comes out dry. That's because water could rise up during the rainy season if their home is in a low-lying area, and exert hydrostatic pressure on the walls resulting in leaks and cracks in your home's foundation.

This is especially probable if their house is built on clay, or if they have to excavate through hard rock or shale to create space for the basement. That's because excessive rain could cause water to perch around it temporarily, inevitably increasing the risk of mould-inducing dampness.

Artificial Causes

Basement contractors dig out a larger area than they require, so they can work on both sides of the walls as they damp or waterproof them. When they are finished, they backfill the spaces on the outside of the basement, usually with the same material they previously excavated.

This creates a leaky fill especially if they're using rock or shale that water filters through more rapidly that the original underlying surface can absorb. This in turn creates a dam of wet material right around the outside basement walls. This can stand for days or even weeks in winter. Just imagine the hydrostatic pressure if you filled the space with water, and you'll have a good idea of the consequences.

Watertable control

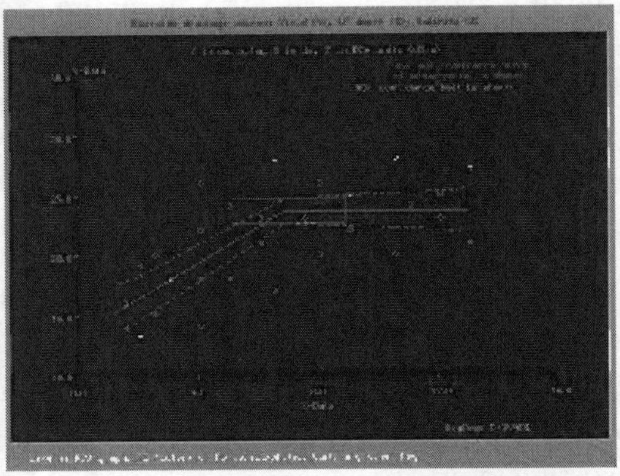

Crop yield (Y) and depth of water table (X in dm)

Watertable control is the practice of controlling the height of the water table by drainage. Its main applications are in agricultural land (to improve the crop yield using agricultural drainage systems) and in cities to manage the extensive underground infrastructure that includes the foundations of large buildings, underground transitsystems, and extensive utilities (water supply networks, sewerage, storm drains, and underground electrical grids).

Purpose

In agricultural land drainage, the purpose of water table control is to establish a depth of the water table that does no longer interfere negatively with the necessary farm operations and crop yields.

In addition, land drainage can help with soil salinity control.

The soil's hydraulic conductivity plays an important role in drainage design. The development of agricultural drainage criteria is required to give the designer and manager of the drainage system a target to achieve in terms of maintenance of an optimum depth of the water table.

Optimization

Optimization of the depth of the water table is related the benefits and costs of the drainage system. The shallower the permissible depth of the water table, the lower the cost of the drainage system to be installed to achieve this depth. However, the lowering of the originally too shallow depth by land drainage entails side effects. These have also to be taken into account, including the costs of mitigation of negative side effects.

The optimization of drainage design and the development of drainage criteriaare discussed in the article on drainage research.

Figure 4 shows an example of the effect of drain depth on soil salinity and various irrigation/drainage parameters as simulated by the SaltModprogram.

Drainage design

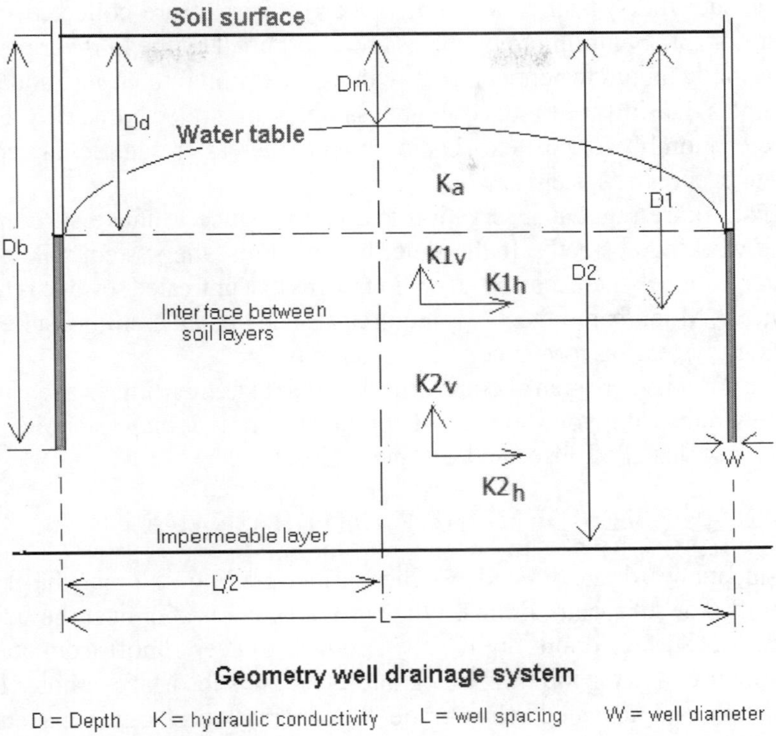

Geometry of a well drainage system

The design of subsurface drainage systems in terms layout, depth and spacing of the drains is often done using subsurface drainage equations with parameters like drain depth, depth of the water table, soil depth, hydraulic conductivity of the soil and drain discharge. The drain discharge is found from an agricultural water balance.

The computations can be done using computer models like EnDrain, which uses the hydraulic equivalent of Joule's law in electricity.

System Design

An RSF was placed between the existing septic tank and trenches (Figure 1). Wastewater moves from the house into a septic tank where solids settle out and some organic matter is decomposed (Figure 2). Liquid effluent moves to the recirculation tank, and from there to the sand filter. The effluent is pumped repeatedly through the sand filter and then flows out via gravity overflow to the drainfield. The system is designed to recirculate effluent through the filter at least five times before it is discharged into the trenches for final treatment.

When wastewater enters the system from the house, an equal amount is delivered into the trenches. A gravity-overflow system ensures that the recirculating tank does not drain completely, which would reduce treatment efficiency and possibly allow the system to freeze during the winter.

To ensure the system is working properly, samples are collected from the recirculation tank. Sampling frequencies are determined by the LGU or permitting agency. Usually, homeowners contract with a professional to collect and analyze these samples. During the first six months samples are collected weekly. Samples are collected monthly for the next six months and, thereafter, samples are collected annually to ensure the system is working properly.

RSFs require regular observation and maintenance to make sure the timer control is working correctly. If the timer control fails, the system will not treat wastewater, although there may be no obvious signs that untreated sewage is reaching the drainfield. Monthly monitoring is required at start-up, but monitoring frequency decreases as system performance is documented.

If you are considering an alternative system, it is extremely important to estimate typical flow rates from your household and make sure that the system is properly sized, designed, installed, inspected, and maintained.

RESIDENTIAL YARD DRAINAGE

Residential yards are drained by either surface grading -- sloping the ground to direct runoff -- or subsurface drains. When groundwater is at or near the surface of the ground, subsurface drains are usually installed to keep runoff from collecting around structures and topsoil from being saturated. For subsurface drains, pipes are buried in channels excavated below grade. Pipes carry excess water to a municipal ditch or drain, or to open space.

Drainage Pipes

Pipes used for subsurface drainage are typically 4 inches in diameter, and solid or perforated plastic, or clay pipe sections. Plastic pipes are made from polyvinyl chloride or polyethylene, either rigid or corrugated, punctured with small openings to capture water draining into the soil. Solid pipes are also used for drainage, where it's desirable for water to be quickly conveyed away. For example, a solid pipe may connect to a downspout to drain roof runoff a safe distance from the foundation.

Underground Channels

The pipes are laid on the gravel-lined underground channel, protected with filter cloth to keep sediment from clogging the pores; concrete pipe sections are placed with a gap between each that allows water in. The channel is sloped slightly to allow runoff to drain. In perforated pipes, the water releases slowly through the pores, and infiltrates into the ground, with excess water carried away to an acceptable discharge point.

Installation and Maintenance

Building codes usually require drains at the base of footings and window wells, and your drainage shouldn't impact your neighbor's property. The pipes should be monitored to make sure do not clog with sediment and debris, and they may need to be cleaned with a sewer snake periodically. Concrete and rigid plastic pipes are stronger than corrugated pipes, which can be damaged by the weight of backfill or the movement of sewer snakes.

Water table control structures

In areas with open canal drainage systems (with or without subsurface drainage), flashboard riser control structures may be used. These are installed in the ditches, and by either inserting or removing a flashboard, the water level in the ditch is either raised or lowered. These structures can also be used to store runoff in canals, which can later be used for sub-irrigation. Surface runoff from drained fields can also be controlled with the aid of the flashboard risers.

Discharge from a subsurface drainage system can be regulated with a PVC control structure similar to that shown in Figure 4. This adjustable water table control structure, attached to the collector outlet, has a vertical riser through which water can be pumped for sub-irrigation.

With pumped drainage outlets, the pumping system can be designed and managed to maintain a desired water level in the sumps, and thus in the field. It is possible to automate the system using a water level sensor installed in a dipwell at the midpoint between subsurface drain laterals (Fouss et al., 1990). The sensor can switch the pump on or off.

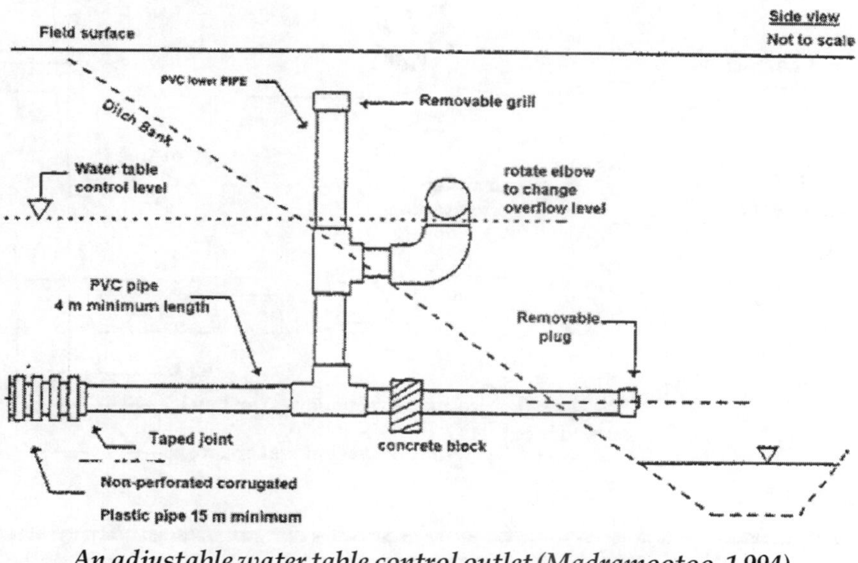

An adjustable water table control outlet (Madramootoo, 1994)

There are two forms of water table management: controlled drainage and sub-irrigation. Controlled drainage restricts the discharge from a subsurface drain outlet or open main drain, resulting in a higher field water table. The water table drops naturally over time due to evaporation and deep seepage, and is only raised if there is precipitation. With sub-irrigation, water is pumped slowly and nearly continually into open ditches or a subsurface drainage system to maintain a near constant water table. When large rainfalls occur and the field water table rises above the desired level, the irrigation pump is stopped. The excess water then drains from a control structure in the ditch or drain outlet. Such systems are shown in Figure.

As a result of higher water tables, water is provided for plant uptake through capillary rise or upward flux. This water helps to meet plant transpiration requirements and reduce drought stress. Corn and soybean yield increases of 10-15% have resulted from sub-irrigation in eastern Canada (Madramootoo et al., 1993; Madramootoo et al., 1995). An added benefit of the higher water table is that it enhances denitrification, thereby reducing nitrate leaching. Denitrification is a microbial process whereby nitrate, rather than oxygen, is used to reduce NO3- to N and NOx. Nitrates retained in the soil profile are available for plant use. Several researchers, including Willardson et al. (1972), Bengtson et al. (1984), Gilliam and Skaggs (1986), Evans et al. (1992) and Lalonde et al. (1995), have shown that water table management can reduce NO3-N in drainage water by over 60%. Denitrification also occurs in water stored in ditches. Other benefits of water table management include: reduced downstream peak flows and drainage volumes, and conservation of soil water to supplement irrigation water requirements. There are, however, some instances where water table management may increase peak flows (Konyha et al.,1992).

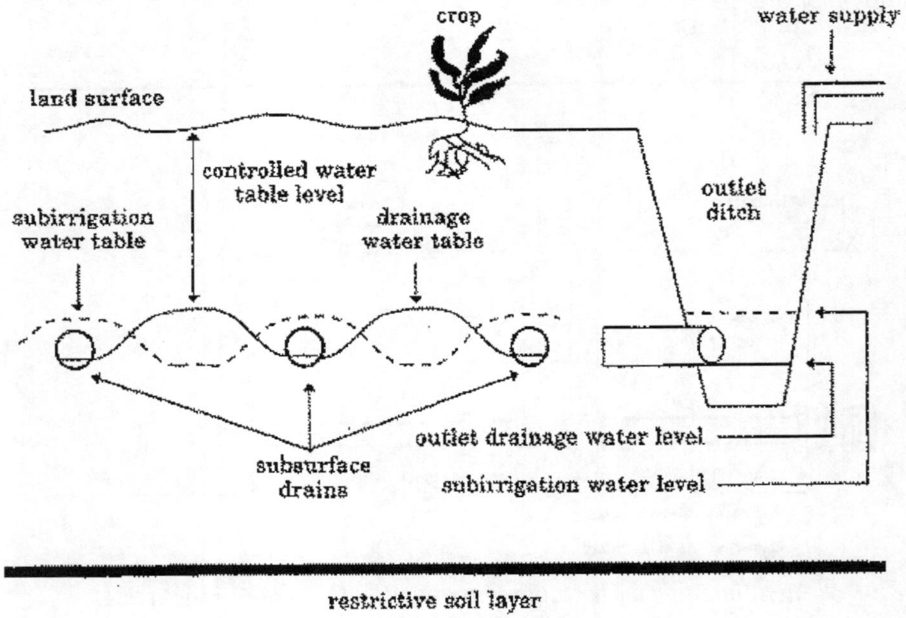

Water table management systems

DRAINMOD model

The DRAINMOD model is one of the most widely-applied models in subsurface drainage system design (Skaggs, 1976, 1978, 1980; Wang et al., 2006; FAO, 2007).

According to Skaggs (1978), the model uses functional algorithms to approximate the hydrological components in soils with shallow water tables. Inputs to the model are weather data, soil data and crop information (effective root zone depth), while its outputs are daily water table depth, drainage discharge, infiltration and runoff. These outputs are primarily estimated from the water balance of a unit soil section located mid-way between 2 drains as per Eq. 1 below:

$$\Delta Va = D + ET + DS - F \tag{1}$$

Where ?Va is the change in water pore space (cm) at any time increment ?t (h); F is the amount of water flowing into the unit soil section as infiltration (cm); D is the amount of water flowing out of the soil in the form of drainage (cm), ET is evapotranspiration (cm) and DS is deep seepage (cm).

Daily water table depths at different drain spacing are computed from the modified steady state Hooghoudt equation (Hooghoudt, 1940):

$$q = \frac{8K_{sat2} d_e h + 4K_{sat1} h^2}{L^2} \tag{2}$$

(m); Ksat1 and Ksat2 are the saturated soil hydraulic conductivities (m'"day-1) for soil layers above and below the drains, respectively; de is the equivalent depth (m); and h is the hydraulic head mid-way between two drains (m) (Oosterbaan, 1975; Fipps and Skaggs, 1989) (see Fig.). According to Oosterbaan (1975), de is a function of the depth to impermeable layer (Dil), drain depth, drain spacing and drain pipe radius

The DRAINMOD model was chosen in this study because it has been tested under a wide range of climatic, crop and soil conditions. For instance, results of the DRAINMOD model performance, in Israel (Sanai and Jain, 2006), Iowa (Singh et al., 2006), South-eastern Purdue Agricultural Center (SEPAC), USA (Wang et al., 2006), Virginia, USA (Mc Mahon et al., (1988), Canada (Dayyani et al.,2009 and Schukla et al., 1994), Italy (Bixio and Bortolini, 1997), and North Carolina, USA (Skaggs 1982), confirm that the DRAINMOD model can reliably mimic subsurface drainage systems under a wide range of soil types and climatic conditions.

9

Rain Water Harvesting Method

Because of the increased propensity for extreme rainfall events in Papua New Guinea (PNG), as well as droughts, the types of water harvesting technique that may be practiced here are more limited than those suitable for drier tropical regions. For instance, semi-permanent water harvesting structures such clay or rock-built bunds, although ideal for harvesting rainwater runoff during normal dry seasons, may create unwelcome water logged conditions during subsequent wet seasons. As such, only temporary water harvesting methods are suitable in this region.

In parts of the Highlands, sweet potato and other vegetable and cereal crops (eg maize) are grown on mounds. One reason for using mounds is to prevent water logging in the crop rooting zone. However, during drought events, crops planted on these raised structures are vulnerable to dehydration.

A means of enhancing moisture availability to mound-grown crops would be to erect temporary barriers between the mounds to trap rainwater runoff in the period prior to a drought, as well as any precipitation during the drought period.

Another method of enhancing moisture availability for mound-grown crops would be to create a hollow on the top of each mound so that rainwater could collect here and percolate down into the crop rooting zone. If necessary, such hollows could be refilled once the rainy season commences to prevent water logging. Having such hollows on the tops of mounds would be beneficial when watering or irrigating the crops during dry periods.

Alternatively, when a drought is expected, rather than planting crops on raised mounds, they could be planted in flat beds or even in shallow depressions on the ground where dew and rainwater could collect and provide much needed moisture. This type of cultivation would also be beneficial where irrigation is employed since water could easily be directed into the crop rooting zone. Where tree crops such as banana are concerned, shallow ditches may be made around each plant to collect rainwater or dew from their stems which subsequently can percolate into the rooting zone.

Rainwater harvesting is the accumulation and deposition of rainwater for reuse on-site, rather than allowing it to run off. Rainwater can be collected from rivers or roofs, and in many places the water collected is redirected to a deep pit (well, shaft, or borehole), a reservoir with percolation, or collected from dew or fog with nets or other tools. Its uses include water for gardens, livestock, irrigation, domestic use with proper treatment, and indoor heating for houses etc. The harvested water can also be used as drinking water, longer-term storage and for other purposes such as groundwater recharge.

RAIN WATER HARVESTING BY FRESHWATER FLOODED FORESTS

Rainwater harvesting is possible by growing fresh water flooded forests without losing the income from the used /submerged land. The main purpose of the rainwater harvesting is to utilize the locally available rainwater to meet water requirements throughout the year without the need of huge capital expenditure. This would facilitate the availability of uncontaminated water for domestic, industrial and irrigation needs.

New approaches

Instead of using the roof for catchment, the RainSaucer, which looks like an upside-down umbrella, collects rain straight from the sky. This decreases the potential for contamination and makes potable water for developing countries a potential application. Other applications of this free standing rainwater collection approach are sustainable gardening and small plot farming.

A Dutch invention called the Groasis Waterboxx is also useful for growing trees with harvested and stored dew and rainwater.

Traditionally, storm water management using detention basins served a single purpose. However, Optimized Real-Time Control (OptiRTC) lets this infrastructure double as a source of rainwater harvesting without compromising the existing detention capacity. This has been used in the EPA headquarters to evacuate stored water prior to storm events, thus reducing wet weather flow while ensuring water availability for later reuse. This has the benefit of increasing water quality released and decreasing the volume of water released during combined sewer overflow events.

Generally, check dams are constructed across the streams to enhance the percolation of surface water into the sub soil strata. The water percolation in the water impounded area of the check dams, can be enhanced artificially many folds by loosening the sub soil strata / overburden by using ANFO explosives as used in open cast mining. Thus local aquifers can be recharged quickly by using the available surface water fully for using in the dry season.

Current use

- In China and Brazil rooftop rainwater harvesting is being practised for providing drinking water, domestic water, water for livestock, water for small irrigation and a way to replenish groundwater levels. Gansu province in China and semi-arid north east Brazil have the largest rooftop rainwater harvesting projects ongoing.
- In Bermuda, the law requires all new construction to include rainwater harvesting adequate for the residents.
- The U.S. Virgin Islands have a similar law.
- In Senegal and Guinea-Bissau, the houses of the Diola-people are frequently equipped with homebrew rainwater harvesters made from local, organic materials.
- In the Irrawaddy Delta of Myanmar, the groundwater is saline and communities rely on mud-lined rainwater ponds to meet their drinking water needs throughout the dry season. Some of these ponds are centuries old and are treated with great reverence and respect.
- In the United States: until 2009 in Colorado, water rights laws almost completely restricted rainwater harvesting; a property owner who captured rainwater was deemed to be stealing it from those who have rights to take water from the watershed. Now, residential well owners that meet certain criteria may obtain a permit to install a *rooftop precipitation collection system* (SB 09-080). Up to 10 large scale pilot studies may also be permitted (HB 09-1129). The main factor in persuading the Colorado Legislature to change the law was a 2007 study that found that in an average year, 97% of the precipitation that fell in Douglas County, in the southern suburbs of Denver, never reached a stream—it was used by plants or evaporated on the ground. In Colorado you cannot even drill a water well unless you have at least 35 acres (14 ha). In New Mexico, rainwater catchment is mandatory for new dwellings in Santa Fe. Texas offers a sales tax exemption on the purchase of rainwater harvesting equipment. Both Texas and Ohio allow the practice even for potable purposes. Oklahoma passed the Water for 2060 Act in 2012, to promote pilot projects for rainwater and graywater use among other water saving techniques.
- In Beijing, some housing societies are now adding rainwater in their main water sources after proper treatment.
- In Ireland, Professor Micheal Mcginley established a project to design a rainwater harvesting prototype in the Biosystems design Challenge Module at University College Dublin

Canada

A number of Canadians have started implementing rainwater harvesting systems for use in stormwater reduction, irrigation, laundry, and lavatory plumbing. Substantial reform to Canadian law since the mid-2000s has increased the use of this technology in agricultural, industrial, and residential use; but ambiguity remains amongst legislation in many provinces. Bylaws and local municipal codes often regulate rainwater harvesting.

India

- **Pan India Potential:** No authenticated potential of rainwater harvesting has been assessed in India. On absolute local basis it has been carried out at some places.
- **Tamil Nadu:** Tamil Nadu was the first state to make Rainwater Harvesting compulsory for every building to avoid groundwater depletion. The scheme was launched in 2001 and has been implemented in all rural areas of Tamil Nadu. We can see posters all over Tamil Nadu including Rural areas to create awareness about harvesting RainwaterTN Govt site. It gave excellent results within five years, and slowly every state took it as a role model. Since its implementation, Chennai saw a 50 percent rise in water level in five years and the water quality significantly improved.
- **Karnataka:** In Bangalore it is mandatory for adoption of rainwater harvesting for every owner or the occupier of a building having the sital area measuring 60 ft (18.3 m) × 40 ft (12. 2 m) and above and for newly constructed building measuring 30 ft (9.1 m) × 40 ft (12.2 m) and above dimension. In this regard, BWSSB has initiated and constructed "Rain Water Harvesting Theme Park" in the name of Sir. M. Visvesvaraya in 1.2 acres (4,900 m) land situated at Jayanagar, Bangalore. In this park, 26 different type of Rain Water Harvesting models are demonstrated along with the water conservation tips. The Auditorium on the first floor is set up with Green Air conditioning system and the same will be utilized to arrange the meeting and showing the video clip about the rainwater harvesting to students as well as general public.
- **Rajasthan:** In Rajasthan, rainwater harvesting has traditionally been practised by the people of the Thar Desert. There are many ancient water harvesting systems in Rajasthan, which have now been revived. Water harvesting systems are widely used in other areas of Rajasthan as well, for example the *chauka* system from the Jaipur district.
- **Kerala:**
- **Maharashtra:** At present, in Pune, rainwater harvesting is compulsory for any new housing society to be registered.
- In Mumbai city in Maharashtra rainwater harvesting is being considered as a good solution to solve the water crisis.

The Mumbai city council is planning to make rainwater harvesting mandatory for large societies. An attempt has been made at the Department of Chemical Engineering, IISc, Bangalore to harvest rainwater using upper surface of a solar still, which was used for water distillation

Israel

The Southwest Center for the Study of Hospital and Healthcare Systems in cooperation with Rotary International is sponsoring a rainwater harvesting model program across the country. The first rainwater catchment system was installed at an elementary school in Lod, Israel. The project is looking to expand to Haifa in its third phase. The Southwest Center has also partnered with the Water Resources Action Project (WRAP) of Washington D.C. WRAP currently has rainwater harvesting projects in the West Bank. Rainwater harvesting systems are being installed in local schools for the purpose of educating schoolchildren about water conservation principles and bridging divides between people of different religious and ethnic backgrounds all while addressing the water scarcity issue that the Middle East faces.

New Zealand

Although New Zealand has plentiful rainfall in the West and South, for much of the country, rainwater harvesting is the normal practice for most rural housing and is encouraged by most councils

Sri Lanka

Rainwater harvesting has been a popular method of obtaining water for agriculture and for drinking purposes in rural homes. The legislation to promote rainwater harvesting was enacted through the Urban Development Authority (Amendment) Act, No. 36 of 2007. Lanka rainwater harvesting forum is leading the Sri Lanka's initiative.

South Africa

The South African Water Research Commission has supported research into rainwater harvesting. Reports on this research are available on their 'Knowledge Hub'. Studies in arid, semi-arid and humid regions have confirmed that techniques such as mulching, pitting, ridging and modified run-on plots are effective for small-scale crop production.

United Kingdom

In the United Kingdom, water butts are often found in domestic gardens and on allotments to collect rainwater, which is then used to water the garden. However, the British government's Code For Sustainable Homes encouraged fitting large underground tanks to new-build homes to collect rainwater for flushing toilets, watering and washing. Ideal designs had the potential to reduce demand on mains water supply by half. The code was revoked in 2015.

United States

- Colorado: Until 2009, water rights laws almost completely restricted rainwater harvesting; a property owner who captured rainwater was deemed to be stealing it from those who have rights to take water from the watershed. Now, residential well owners that meet certain criteria may obtain a permit to install a *rooftop precipitation collection system* (SB 09-080). Up to 10 large scale pilot studies may also be permitted (HB 09-1129). The main factor in persuading the Colorado Legislature to change the law was a 2007 study that found that in an average year, 97% of the precipitation that fell in Douglas County, in the southern suburbs of Denver, never reached a stream—it was used by plants or evaporated on the ground. In Colorado, one is not permitted to drill a water well unless one has at least 35 acres (14 ha) of land.

- New Mexico: In New Mexico, rainwater catchment is mandatory for new dwellings in Santa Fe.
- Texas offers a sales tax exemption on the purchase of rainwater harvesting equipment. Both Texas and Ohio allow the practice even for potable purposes.
- Oklahoma passed the Water for 2060 Act in 2012, to promote pilot projects for rainwater and graywater use among other water saving techniques.

Non-traditional

- In 1992, American artist Michael Jones McKean created an artwork in Omaha, Nebraska at the Bemis Center for Contemporary Art that created a fully sustainable rainbow in the Omaha skyline. The project collected thousands of gallons of rainwater storing the water in six daisy-chained 12, 000 gallons tanks. The massive logistical undertaking, during its 5-month span, was one of the largest urban rainwater harvesting sites in the American Mid-West.

MOISTURE CONSERVATION

During drought periods or normal dry season, conditions in the dry lowlands, preventing moisture loss from the soil is important. Drought causes water stress and wilting of plant leaves. Low humidity and large temperature changes also affect crop reproduction cycles preventing the formation of seeds and fruit. During drought, soil is exposed to heat and wind and can become very hot during the day and may be blown away as dust. Lack of moisture also causes it to harden and crack enabling pests such as weevils to attack crop roots.

The simplest method for minimising the impacts of high temperature, low humidity and lack of rainfall on cropping areas is to apply mulch. Mulch is a layer of dead vegetation laid on the ground both to control weeds and to minimise moisture loss by evaporation.

Mulch can be spread on gardens after they have been dug and either before or after the crop has been planted. Some crops may be planted through the mulch, whereas for others, it is best to spread the mulch on the soil surface after the crop has started growing.

Mulches shade the soil, lowering surface temperatures and reducing the rate of water loss. They also protect soil from wind erosion and frost. Furthermore, in the wet-up periods following drought, they can reduce soil erosion by sheltering the soil from heavy rain and preventing excess rain water running off fields too quickly.

Cultivation under tree canopies is another option for conserving moisture in cropping systems. The trees shade the crops and the soil, lowering soil temperatures and thus minimising evaporation losses.

In the present scenario management and distribution of water has become centralized. People depend on government system, which has resulted in disruption of community participation in water management and collapse of traditional water harvesting system. As the water crisis continues to become severe, there is a dire need of reform in water management system and revival of traditional systems. Scientific and technological studies needs to be carried out to assess present status so as to suggest suitable mitigative measures for the revival to traditional system/ wisdom. Revival process should necessarily be backed by people's initiative and active public participation.

Living creatures of the universe are made of five basic elements, viz., Earth, Water, Fire, Air and Sky, Obviously, water is one of the most important elements and no creature can survive without it. Despite having a great regard for water, we seem to have failed to address this sector seriously. Human being could not save and conserve water and it sources, probably because of its availability in abundance. But this irresponsible attitude resulted in deterioration of water bodies with respect to quantity and quality both. Now, situation has arrived when even a single drop of water matters. However. " better late than never", we have not realized the seriousness of this issue and initiated efforts to overcome those problems.

System of collection rainwater and conserving for future needs has traditionally been practiced in India. The traditional systems were time-tested wisdom of not only appropriate technology of Rainwater Harvesting, but also water management systems, where conservation of water was the prime concern. Traditional water harvesting systems were Bawaries, step wells, jhiries, lakes, tanks etc. These were the water storage bodies to domestic and irrigation demands. People were themselves responsible for maintenance to water sources and optimal use of water that could fulfill their needs.

What is Rainwater harvesting?

The term rainwater harvesting is being frequently used these days, however, the concept of water harvesting is not new for India. Water harvesting techniques had been evolved and developed centuries ago. Ground water resource gets naturally recharged through percolation. But due to indiscriminate development and rapid urbainzation, exposed surface for soil has been reduced drastically with resultant reduction in percolation of rainwater, thereby depleting ground water resource. Rainwater harvesting is the process of augmenting the natural filtration of rainwater in to the underground formation by some artificial methods. "Conscious collection and storage of rainwater to cater to demands of water, for drinking, domestic purpose & irrigation is termed as Rainwater Harvesting. "

Why harvest rainwater ?

This is perhaps one of the most frequently asked question, as to why one should harvest rainwater. There are many reasons but following are some of the important ones.

- To arrest ground water decline and augment ground water table
- To beneficiate water quality in aquifers
- To conserve surface water runoff during monsoon
- To reduce soil erosion
- To inculcate a culture of water conservation

How to harvest rainwater: Broadly there are two ways of harvesting rainwater:

(i) Surface runoff harvesting

(ii) Roof top rainwater harvesting

Surface runoff harvesting:

In urban area rainwater flows away as surface runoff. This runoff could be caught and used for recharging aquifers by adopting appropriate methods.

Roof top rainwater harvesting (RTRWH):

It is a system of catching rainwater where it falls. In rooftop harvesting, the roof becomes the catchments, and the rainwater is collected from the roof of the house/building. It can either be stored in a tank or diverted to artificial recharge system. This method is less expensive and very effective and if implemented properly helps in augmenting the ground water level of the area.

COMPONENTS OF THE ROOF TOP RAINWATER HARVESTING SYSTEM

The illustrative design of the basic components of roof top rainwater harvesting system is given in the following typical schematic diagram/ The system mainly constitutes of following sub components:

- Catchment
- Transportation

The surface that receives rainfall directly is the catchment of rainwater harvesting system. It may be terrace, courtyard, or paved or unpaved open ground. The terrace may be flat RCC/stone roof or sloping roof. Therefore the catchment is the area, which actually contributes rainwater to the harvesting system.

Transportation

Rainwater from rooftop should be carried through down take water pipes or drains to storage/harvesting system. Water pipes should be UV resistant (ISI HDPE/ PVC pipes) of required capacity. Water from sloping roofs could be caught through gutters and down take pipe. At terraces, mouth of the each drain should have wire mesh to restrict floating material.

First Flush First flush is a device used to flush off the water received in first shower. The first shower of rains needs to be flushed-off to avoid contaminating storable/rechargeable water by the probable contaminants of the atmosphere and the catchment roof. It will also help in cleaning of silt and other material deposited on roof during dry seasons Provisions of first rain separator should be made at outlet of each drainpipe.

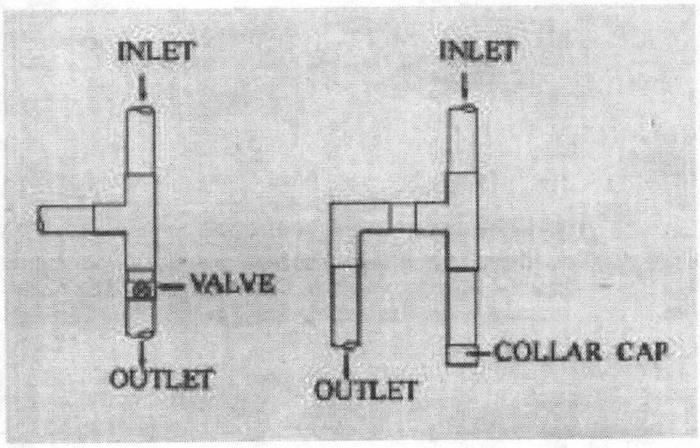

Filter

There is always some skepticism regarding Roof Top Rainwater Harvesting since doubts are raised that rainwater may contaminate groundwater. There is remote possibility of this fear coming true if proper filter mechanism is not adopted. Secondly all care must be taken to see that underground sewer drains are not punctured and no leakage is taking place in close vicinity. Filters are used fro treatment of water to effectively remove turbidity, colour and microorganisms. After first flushing of rainfall, water should pass through filters. There are different types of filters in practice, but basic function is to purify water.

Sand Gravel Filter

These are commonly used filters, constructed by brick masonry and filleted by pebbles, gravel, and sand as shown in the figure. Each layer should be separated by wire mesh.

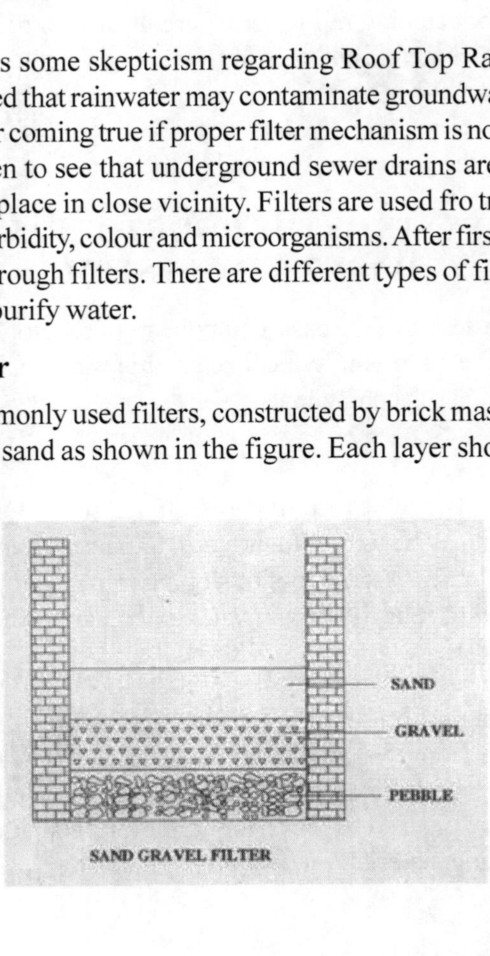

SAND GRAVEL FILTER

Charcoal Filter

Charcoal filter can be made in-situ or in a drum. Pebbles, gravel, sand and charcoal as shown in the figure should fill the drum or chamber. Each layer should be separated by wire mesh. Thin layer of charcoal is used to absorb odor if any.

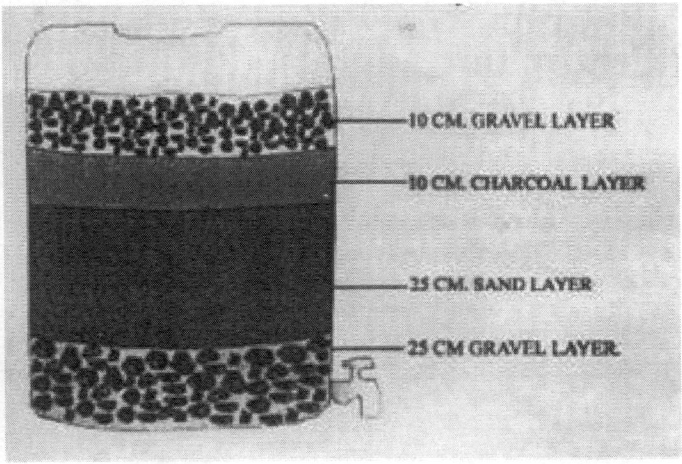

PVC- Pipe filter

This filter can be made by PVC pipe of 1 to 1. 20 m length; Diameter of pipe depends on the area of roof. Six inches dia. pipe is enough for a 1500 Sq. Ft. roof and 8 inches dia. pipe should be used for roofs more then 1500 Sq. Ft. Pipe is divided into three compartments by wire mesh. Each component should be filled with gravel and sand alternatively as shown in the figure. A layer of charcoal could also be inserted between two layers. Both ends of filter should have reduce of required size to connect inlet and outlet. This filter could be placed horizontally or vertically in the system.

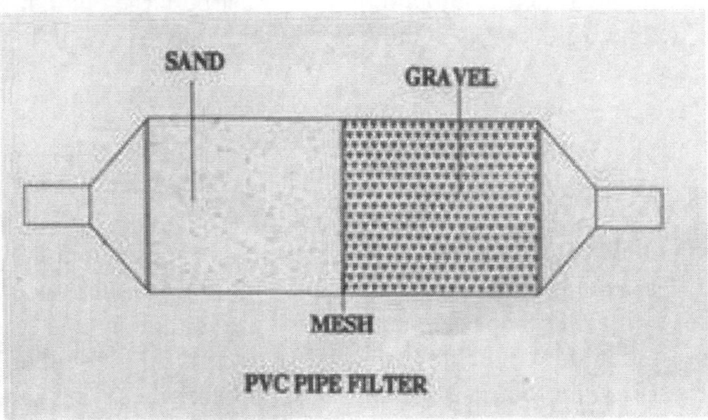

Sponge Filter

It is a simple filter made from PVC drum having a layer of sponge in the middle of drum. It is the easiest and cheapest form filter, suitable for residential units.

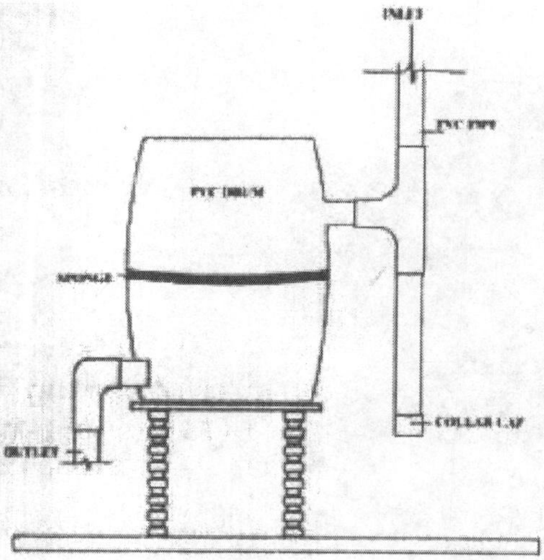

SPONGE FILTER

System setup

Rainwater harvesting systems can range in complexity, from systems that can be installed with minimal skills, to automated systems that require advanced setup and installation. The basic Rainwater harvesting system is more of a plumbing job than a technical job as all the outlets from the building terrace are connected through a pipe to an underground tank that stores water.

Systems are ideally sized to meet the water demand throughout the dry season since it must be big enough to support daily water consumption. Specifically, the rainfall capturing area such as a building roof must be large enough to maintain adequate flow. The water storage tank size should be large enough to contain the captured water.

For low-tech systems, there are many low-tech methods used to capture rainwater: rooftop systems, surface water capture, and pumping the rainwater that has already soaked into the ground or captured in reservoirs and storing it into tanks (cisterns).

Before a rainwater harvesting system is built, it is helpful to use digital tools. For instance, if you want to detect if a region has a high rainwater harvesting potential, rainwater harvesting GIS maps can be made using an online interactive tool. Or if you need to estimate how much water is needed to fulfill a community's water needs, the Rain is Gain tool helps with this. Tools like these can save time and money before a commitment to build a system is undertaken, in addition to making the project sustainable and last a long time.

Life Cycle Assessment: Design for Environment

Contemporary system designs require an analysis of not only the economic and technical performance of a system, but also the environmental performance. Life Cycle Assessment is a methodology used to evaluate the environmental impacts of a precut or systems, from cradle-to-grave of its' lifetime. Devkota et al., developed such a methodology for rainwater harvesting, and found that the building design (e.g., dimensions) and function (e.g., educational, residential, etc.) play critical roles in the environmental performance of the system. The Economic and Environmental Analysis of Sanitations Technologies, EEAST model evaluates the greenhouse gas emissions and cost of such systems over the lifetime of a variety of building types.

To address the functional parameters of rainwater harvesting systems, a new metric was developed - the demand to supply ratio (D/S) - identifying the ideal building design (supply) and function (demand) in regard to the environmental performance of rainwater harvesting for toilet flushing. With the idea that supply of rainwater not only saves the potable water but also saves the stormwater entering the combined sewer network (thereby requiring treatment), the savings in environmental emissions were higher if the buildings are connected to a combined sewer network compared to separate one.

ADVANTAGES

Rainwater harvesting provides an independent water supply during regional water restrictions and in developed countries is often used to supplement the main supply. It provides water when there is a drought, can help mitigate flooding of low-lying areas, and reduces demand on wells which may enable groundwater levels to be sustained. It also helps in the availability of potable water as rainwater is substantially free of salinity and other salts. Application of rainwater harvesting in urban water system provides a substantial benefit for both water supply and wastewater subsystems by reducing the need for clean water in water distribution system, less generated stormwater in sewer system, as well as a reduction in stormwater runoff polluting freshwater bodies. There has been a large body of work focused on the development of Life Cycle Assessment and Life Cycle Costing methodologies to assess the level of environmental impacts and money that can be saved by implementing rainwater harvesting systems.

More development and knowledge is required to understand the benefits rainwater harvesting can provide to agriculture. Many countries especially those with an arid environment use rainwater harvesting as a cheap and reliable source of clean water. To enhance irrigation in arid environments, ridges of soil are constructed in order to trap and prevent rainwater from running down hills and slopes. Even in periods of low rainfall, enough water is collected in order for crops to grow. Water can be collected from roofs, dams, and ponds can be constructed in order to hold large quantities of rainwater so that even on days where there is little to no rainfall, there is enough available to irrigate crops.

Quality

The concentration of contaminants is reduced significantly by diverting the initial flow of run-off water to waste. Improved water quality can also be obtained by using a floating draw-off mechanism (rather than from the base of the tank) and by using a series of tanks, withdraw from the last in series. Pre-filtration is a common practice used in the industry to ensure that the water entering the tank is free of large sediment. Pre-filtration is important to keep the system healthy.

Conceptually, a water supply system should match the quality of water with the end use. However, in most of the developed world high quality potable water is used for all end uses. This approach wastes money and energy and imposes unnecessary impacts to the environment. Supplying rainwater that has gone through preliminary filtration measures for non-potable water uses, such as toilet flushing, irrigation, and laundry, may be a significant part of a sustainable water management strategy.

METHODS OF ROOF TOP RAINWATER HARVESTING

Storage of Direct use

In this method rain water collected from the roof of the building is diverted to a storage tank. The storage tank has to be designed according to the water requirements, rainfall and catchment availability. Each drainpipe should have mesh filter at mouth and first flush device followed by filtration system before connecting to the storage tank. It is advisable that each tank should have excess water over flow system.

Excess water could be diverted to recharge system. Water from storage tank can be used for secondary purposes such as washing and gardening etc. This is the most cost effective way of rainwater harvesting. The main advantage of collecting and using the rainwater during rainy season is not only to save water from conventional sources, but also to save energy incurred on transportation and distribution of water at the doorstep. This also conserve groundwater, if it is being extracted to meet the demand when rains are on.

Recharging ground water aquifers

Ground water aquifers can be recharged by various kinds of structures to ensure percolation of rainwater in the ground instead of draining away from the surface. Commonly used recharging methods are:-

a) Recharging of bore wells

b) Recharging of dug wells.

c) Recharge pits

d) Recharge Trenches
e) Soak ways or Recharge Shafts
f) Percolation Tanks

Recharging of bore wells

Rainwater collected from rooftop of the building is diverted through drainpipes to settlement or filtration tank. After settlement filtered water is diverted to bore wells to recharge deep aquifers. Abandoned bore wells can also be used for recharge.

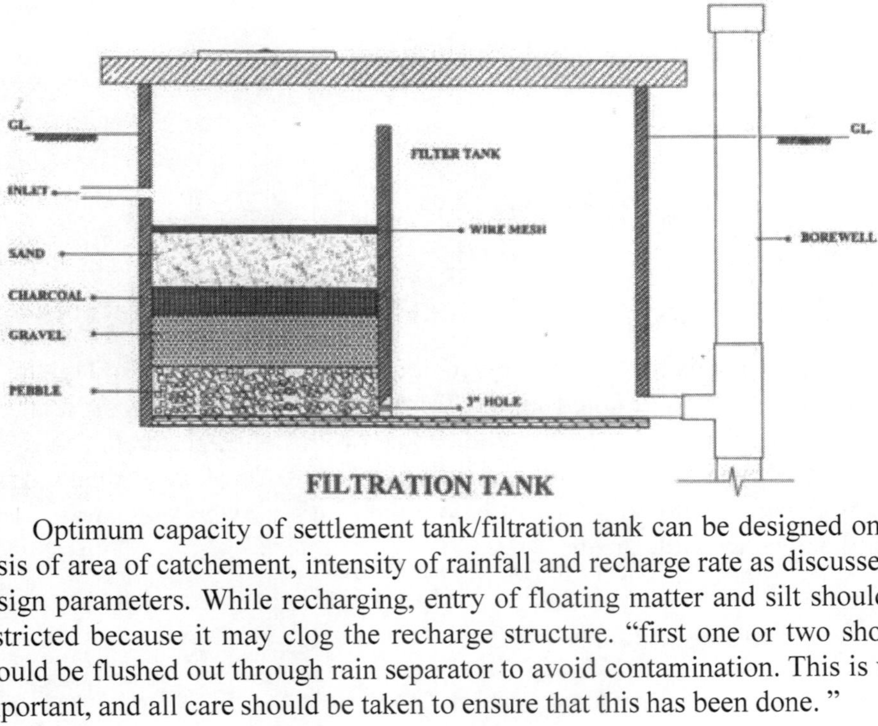

FILTRATION TANK

Optimum capacity of settlement tank/filtration tank can be designed on the basis of area of catchement, intensity of rainfall and recharge rate as discussed in design parameters. While recharging, entry of floating matter and silt should be restricted because it may clog the recharge structure. "first one or two shower should be flushed out through rain separator to avoid contamination. This is very important, and all care should be taken to ensure that this has been done. "

Recharge Pits

Recharge pits are small pits of any shape rectangular, square or circular, contracted with brick or stone masonry wall with weep hole at regular intervals. to of pit can be covered with perforated covers. Bottom of pit should be filled with filter media.

The capacity of the pit can be designed on the basis of catchment area, rainfall intensity and recharge rate of soil. Usually the dimensions of the pit may be of 1 to 2 m width and 2 to 3 m deep depending on the depth of pervious strata. These pits are suitable for recharging of shallow aquifers, and small houses.

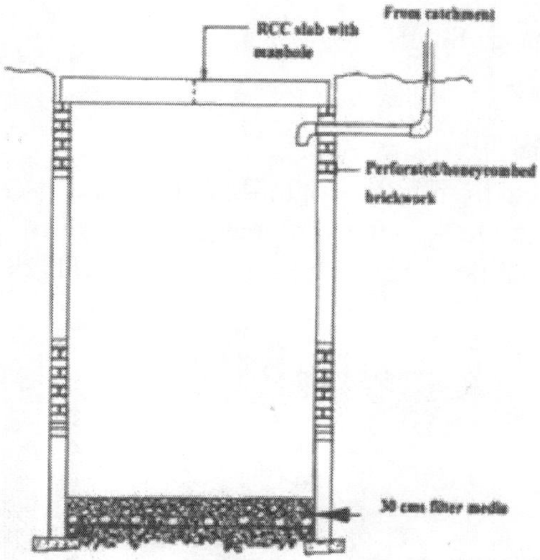

Typical recharge pit

Soak away or Recharge Shafts

Soak away or recharge shafts are provided where upper layer of soil is alluvial or less pervious. These are bored hole of 30 cm dia. up to 10 to 15 m deep, depending on depth of pervious layer. Bore should be lined with slotted/perforated PVC/MS pipe to prevent collapse of the vertical sides. At the top of soak away required size sump is constructed to retain runoff before the filters through soak away. Sump should be filled with filter media.

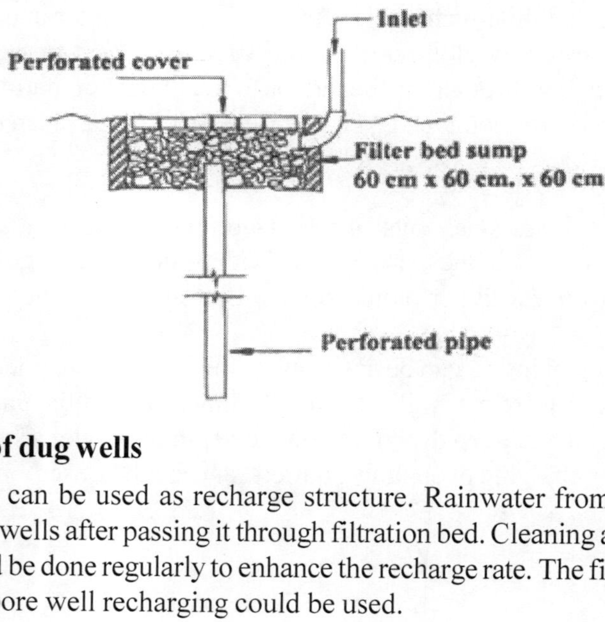

Recharging of dug wells

Dug well can be used as recharge structure. Rainwater from the rooftop is diverted to dug wells after passing it through filtration bed. Cleaning and desalting of dug well should be done regularly to enhance the recharge rate. The filtration method suggested for bore well recharging could be used.

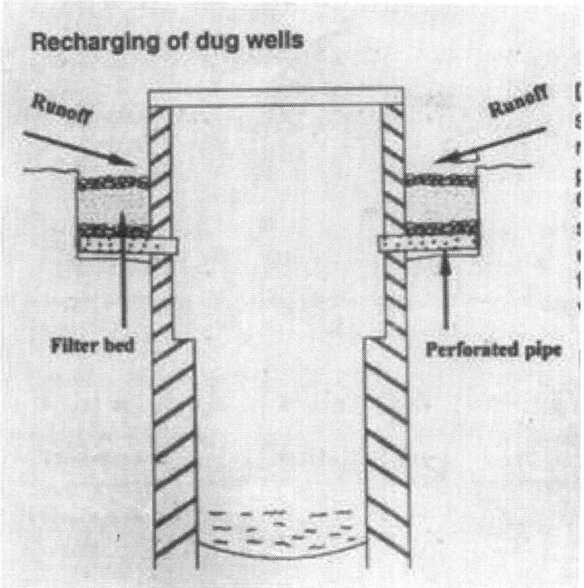

Recharge Trenches

Recharge trench in provided where upper impervious layer of soil is shallow. It is a trench excavated on the ground and refilled with porous media like pebbles, boulder or brickbats. it is usually made for harvesting the surface runoff. Bore wells can also be provided inside the trench as recharge shafts to enhance percolation. The length of the trench is decided as per the amount of runoff expected. This method is suitable for small houses, playgrounds, parks and roadside drains. The recharge trench can be of size 0.50 to 1.0 m wide and 1.0 to 1. 5 m deep.

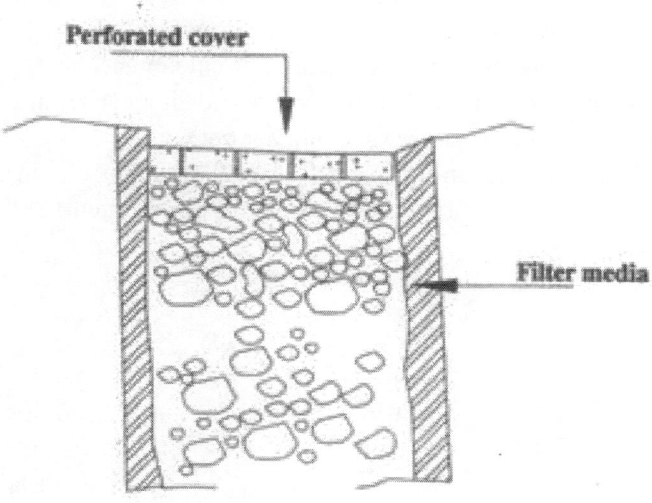

Recharge Trenches

Percolation tanks

Percolation tanks are artificially created surface water bodies, submerging a land area with adequate permeability to facilitate sufficient percolation to recharge the ground water. These can be built in big campuses where land is available and topography is suitable.

Surface run-off and roof top water can be diverted to this tank. Water accumulating in the tank percolates in the solid to augment the ground water. The stored water can be used directly for gardening and raw use. Percolation tanks should be built in gardens, open spaces and roadside green belts of urban area.

Do's and Don'ts

Harvested rainwater is used for direct usage or for recharging aquifers. It is most important to ensure that the rainwater caught is free from pollutants. Following precautionary measures should be taken while harvesting rainwater:-

- Roof or terraces uses for harvesting should be clean, free from dust, algal plants etc.
- Roof should not be painted since most paints contain toxic substances and may peel off.
- Do not store chemicals, rusting iron, manure or detergent on the roof.
- Nesting of birds on the roof should be prevented.
- Terraces should not be used for toilets either by human beings or by pets.
- Provide gratings at mouth of each drainpipe on terraces to trap leaves debris and floating materials.
- Provision of first rain separator should be made to flush off first rains.
- Do not use polluted water to recharge ground water.
- Ground water should only be recharged by rainwater.
- Before recharging, suitable arrangements of filtering should be provided.
- Filter media should be cleaned before every monsoon season.
- During rainy season, the whole system (roof catchment, pipes, screens, first flush, filters, tanks) should be checked before and after each rain and preferably cleaned after every dry period exceeding a month.

At the end of the dry season and just before the first shower of rain is anticipated, the storage tank should be scrubbed and flushed off all sediments and debris.

Bibliography

- Adams, J.A. 1986. Dirt. College Station, Texas : Texas A&M University Press ISBN 0-89096-301-0
- Agricultural Drainage Criteria. Chapter 17 in: H.P.Ritzema (ed. 1994), Drainage Principles and Applications, ILRI Publication 16, p.635-690. International Institute for Land Reclamation and Improvement (ILRI), Wageningen, The Netherlands. ISBN 90 70754 3 39. Free download from : , under nr. 1, or directly as PDF :
- Birkeland, Peter W. Soils and Geomorphology, 3rd Edition. New York: Oxford University Press, 1999.
- Buol, S. W.; Hole, F. D. and McCracken, R. J. (1973), Soil Genesis and Classification (First ed.), Ames, IA: Iowa State University Press, ISBN 0-8138-1460-X.
- Buol, S.W. (1990), Soil genesis and classification, Ames, Iowe: Iowa State University Press, pp. 36, doi:10.1081/E-ESS, ISBN 0813828732
- Chesworth, Edited by Ward (2008), Encyclopedia of soil science, Dordrecht, Netherland: Springer, xxiv, ISBN 1402039948
- Copley, Jon (August 25, 2005). "Millions of bacterial species revealed underfoot". Reed Business Information Ltd. New Scientist. Retrieved 19 April 2010.
- Dan (2000), Ecology and management of forest soils, New York: John Wiley, pp. 88-92, ISBN 0471194263
- De Jong, M.H. 1979. Drainage of structured clay soils. In: J. Wesseling (ed.), Proceedings of the International Drainage Workshop. ILRI Publication 25, pp. 268-280. International Institute for Land Reclamation and Improvement (ILRI), Wageningen, The Netherlands. ISBN 90-70260-54-9
- Drainage Manual: A Guide to Integrating Plant, Soil, and Water Relationships for Drainage of Irrigated Lands, Interior Dept., Bureau of Reclamation, 1993, ISBN 0-16-061623-9
- Faulkner, Edward H. Plowman's Folly. New York, Grosset & Dunlap. 1943. ISBN 0-933280-51-3

- General case studies of drainage methods and systems in agricultural land. Lecture note, International Course on Land Drainage (ICLD), International Institute for Land Reclamation and Improvement (ILRI), Wageningen, The Netherlands. Download from : , under nr. 4, or directly as PDF :
- J.Wijdieks and M.G.Bos, 1994. Pumps and Pumping Stations. In: H.P.Ritzema (ed.), Drainage Principles and Applications, ILRI Publication 16, p.965-1000. International Institute for Land Reclamation and Improvement (ILRI), Wageningen, The Netherlands. ISBN 90 70754 3 39.
- James A. Danoff-Burg, Columbia University The Terrestrial Influence: Geology and Soils.
- Jenny, Hans, Factors of Soil Formation: A System of Quantitative Pedology 1941.
- Johnson, D.L.; Ambrose, S.H.; Bassett, T.J.; Bowen, M.L.; Crummey, D.E.; Isaacson, J.S.; Johnson, D.N.; Lamb, P. et al. (1997). "Meanings of environmental terms". Journal of Environmental Quality 26: 581-589.
- Kohne, John Maximilian; Koehne Sigrid, Simunek Jirka (2009), "A review of model applications for structured soils: a) Water flow and tracer transport", Journal of Contaminant Hydrology 104 (1-4): 4-35, doi:10.1016/j.jconhyd.2008.10.002, PMID 19012994
- Logan, W. B., Dirt: The ecstatic skin of the earth. 1995 ISBN 1-57322-004-3
- M.G.Bos, 1994. Subsurface Drainage Systems. In: H.P.Ritzema (ed.), Drainage Principles and Applications, ILRI Publication 16, p.725-798. International Institute for Land Reclamation and Improvement (ILRI), Wageningen, The Netherlands. ISBN 90 70754 3 39
- Nosenko, P.P. and I.S. Zonn 1976. Land Drainage in the World. ICID Bulletin 25, 1, pp.65-70.
- Oosterbaan, R.J., H.A. Gunneweg and A. Huizing 1987. Water control for rice cultivation in small valleys of West Africa. In: ILRI Annual Report 1986 pp. 30-49. International Institute for Land Reclamation and Improvement (ILRI), Wageningen, The Netherlands. Download from : , under nr. 8, or directly as PDF :
- Qorani, M., M.S. Abdel Dayem and R.J. Oosterbaan 1990. Evaluation of restricted subsurface drainage in rice fields. In: Symposium on land drainage for salinity control in arid and semi arid regions, Vol. 3. Drainage Research Institute, Cairo, pp. 415-423
- R. B. Brown (September 2003). "Soil Texture". Fact Sheet SL-29. University of Florida, Institute of Food and Agricultural Sciences.
- R. J. Edwards (1998-02-15). "Typical Soil Characteristics of Various Terrains".
- R.J.Sevenhuijsen, 1994. Surface Drainage Systems. In: H.P.Ritzema (ed.), Drainage Principles and Applications, ILRI Publication 16, p.799-826. International Institute for Land Reclamation and Improvement (ILRI), Wageningen, The Netherlands. ISBN 90 70754 3 39

Index

E

F

G

H

I

K

L